CENELEC 50128 and IEC 62279 Standards

CENELEC 50128 and IEC 62279 Standards

Jean-Louis Boulanger

iSTE

WILEY

First published 2015 in Great Britain and the United States by ISTE Ltd and John Wiley & Sons, Inc.

ISTE Ltd
27-37 St George's Road
London SW19 4EU
UK

www.iste.co.uk

John Wiley & Sons, Inc.
111 River Street
Hoboken, NJ 07030
USA

www.wiley.com

Library of Congress Control Number: 2015931031

British Library Cataloguing-in-Publication Data
A CIP record for this book is available from the British Library
ISBN 978-1-84821-634-1

Contents

Introduction

I.1. Objective

Railways are subject to both normative and legal frameworks (laws, decrees, regulations, etc.) which differ from one country to another. At the European level, the legal framework includes European and national texts. It should be noted that this normative and legislatory framework is fairly new (the earliest published standards date from the mid-1990s, and the earliest laws passed from 2004). Figure I.1 presents the main standards which apply to the building of a railway system.

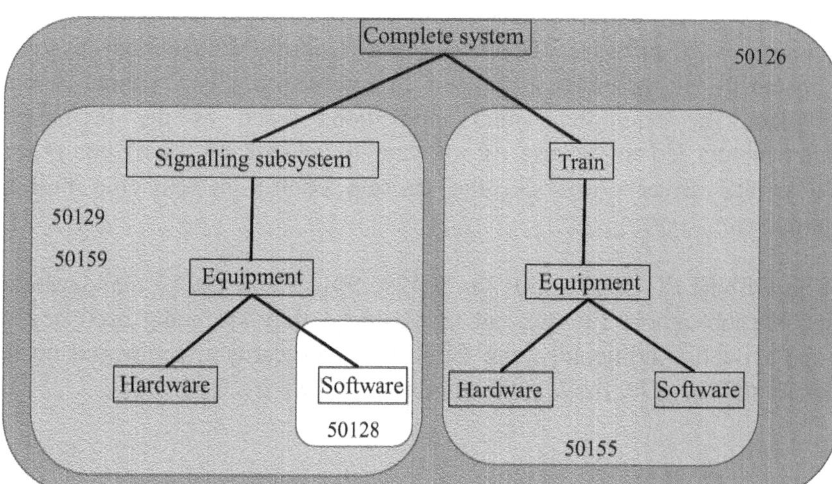

Figure I.1. *Normative context*

As Figure I.1 illustrates, the domain of railways is divided into two parts: applications relating to signaling and applications onboard the trains. In fact, it is necessary to add a third family of applications – "Miscellaneous": this category would include means of energy management, the systems that run travelators and escalators, information systems, applications for management of the auxiliary systems (e.g. tunnel ventilation, fire detection, etc.) – indeed anything at all which can be connected to the railway system. The auxiliary systems are no less important than the primary ones. The fire-detection and tunnel-ventilation system is a system connected to the domain which could prevent a tunnel from filling with smoke in an evacuation situation. Thus, this system has a bearing on safety.

The subdivision of the normative framework stems from the fact that, originally, safety in railway systems was based on signaling (changing the state of the signals depending on the presence or absence of trains on the track), and the train driver was responsible for respecting the commands shown to him by the signals.

When software was first used, it rendered the principles of signaling more flexible (the transmission of signaling information to the driver's cab as a report, virtual division of the track, etc.). The second step involved installing software on board the trains to handle non-safety-related information and to develop specific, small functions. The need to keep weight and costs under control led manufacturers to replace the copper hardwired systems and relays with software (TCMS – Train Control/Management Systems, for instance). In addition, the need to evolve quickly (without having to replace the equipment) and innovations (such as the permanent-magnet motor) led to the use of software for classic pieces of equipment, such as the driver's joystick, the traction control system, the braking system, etc.

The CENELEC 50126, 50128, 50129, 50159 and 50155 standards are applicable throughout Europe, but increasingly, they are being used beyond Europe as well. Additionally, the CENELEC standards are mirrored on the international scene by the IEC standards[1], as shown by Table I.1.

1 "IEC" stands for *International Electrotechnical Commission*. For further information, see: www.iec.ch/.

This book presents the 2011 version of CENELEC[2] 50128 standard [CEN 11a] and its implementation. In Chapter 13, we shall give a detailed breakdown of the differences between CENELEC 50128:2011 and its IEC equivalent: 62279.

CENELEC	IEC	Comments
50126:1999	62278:2002	Same
50129:2003	62425:2007	Same
50128:2001	62279:2002	Identical with the exception of the first page
50128:2011	62279:2014-draft	The IEC draft contains notable differences in relation to the CENELEC document (e.g. additional constraints for certification of tools, etc.)
50159-1 50159-2	62280-1:2002 62280-2:2002	Same
50159:2011	62280:2014	Same
50155	60571	Identical, except for the fact that the IEC standard contains additional explanations

Table I.1. *Breakdown of CENELEC and IEC standards*

CENELEC 50128:2011 identifies a process for creating software for railway applications, and identifies the resources which need to be mobilized in order to achieve the set level of assurance. It introduces new requirements such as separation between the generic software and the settings data, certification of the tools, the need to document and the need to stay abreast of maintenance and the rollout of new versions of the software.

We are going to present this new version of the standard, but above all, we shall give references to the fundamental reading necessary to put the standard into practice.

I.2. Reminder

Safety of railway applications was originally based on the management of signaling. With automated systems such as the metro (see Line 14 of the

2 "CENELEC" stands for *Comité Européen de Normalisation ELECtrotechnique* (European Commission for Electrotechnical Standardization). For further information, see: www.cenelec.eu/.

Paris metro[3] and/or the VAL (*Véhicule Automatique Léger* – Lightweight Automated Vehicle)[4] in Charles de Gaulle Airport), software is used to enhance safety management. The 2001 version of CENELEC 50128 [CEN 01a] was written to define a context by which to manage the safety of the software used. This version of the standard benefited from the advent of numerous software-based systems.

Figure I.2. *The VAL at CdG Airport, standing at the platform*[5]

Since the release of this version, the use of software has expanded to all parts of the railway industry (driver support, driver joystick, door management, traction management, management of sensor settings, tunnel ventilation management system, etc.), and it has become necessary to take new problems into account, such as maintenance and deployment. The maintenance of software goes above and beyond simple correction of anomalies, to the handling of evolution of a range of equipment, different versions of which may be used by different operators. Hence, it is necessary to take maintenance measures which take account of the versions employed and a rollout process which enables us to guarantee the systems will work properly after new versions are rolled out.

3 The design and approval of the SAET-METEOR (developed by MATRA-transport – now SIEMENS – for the RATP, see [MAT 98]), brought into operation in 1998, greatly contributed to the formulation of the 2001 version 2001 of CENELEC 50128.

4 The first VAL began operating in Lille in 1983. Today, it is used in Taipei and Toulouse, Rennes and Turin (since January 2006). With regard to the rollout of the VAL, at least 119km of track have been laid worldwide, and over 830 carriages are currently in service or under construction. The VAL at CDG combines VAL technology and additional digital equipment based on the B method [ABR 96].

5 Photo taken by Jean-Louis Boulanger.

The creation of a software application is based on people and on the use of complex tools. In relation to the first point, the new version of the standard places emphasis on the management of skills and responsibilities. On the second point, the tools can have an impact on the executable content (code generators, compilers, etc.) and/or on the verification (test environment, tool for checking programming rules, etc.), so it is necessary to qualify and/or certify the tools that are used. It should be noted that this notion of qualification is one which has been introduced in the newly-updated set of standards (IEC 61508 [IEC 08], ISO 26262 [ISO 11], CENELEC 50128 [CEN 11a], etc.).

I.3. Overview

We have given a brief presentation of the CENELEC 50128 standard and have begun to introduce the changes which were made to it in the 2011 version. Thus, in the remaining chapters of this book, we shall present the 2011 version of the CENELEC 50128 standard and the principles of its implementation.

The chapters of this book are presented as follows:

– Chapter 1: software in the system;

– Chapter 2: history of the CENELEC framework and structure of the 50128 standard;

– Chapter 3: definitions in the system and allocation to the software packages;

– Chapter 4: quality assurance on the software (quality management, organization management, checking and validation, etc.). Application of Chapter 6 of the CENELEC 50128 standard;

– Chapter 5: requirement management;

– Chapter 6: the specific application and data-based settings. Implementation of Chapter 8 of the CENELEC 50128 standard;

– Chapter 7: development of the generic application. Implementation of Chapter 7 of the CENELEC 50128 standard;

– Chapter 8: model, modeling and formalization;

– Chapter 9: certification of tools;

– Chapter 10: maintenance and rollout. Implementation of Chapter 9 of the CENELEC 50128 standard;

– Chapter 11: independent evaluation;

– Conclusion.

1

From the System to the Software

1.1. Introduction

The automation of numerous command systems (in railways, the aeronautics, automotive, nuclear industries, etc.) and/or process control systems (production, etc.), and the replacement of logical or analog systems involving little interaction by highly-integrated systems, have led to a considerable expansion of the domain of functional safety, taking account of the features and peculiarities of computer systems.

Dependability relates to applications for which it is crucial to ensure a continuous good level of service (reliability), because human lives are at stake (transport, nuclear energy, etc.), because of the high level of investment which would be lost were the calculation to go wrong (space, chemical production process, etc.), or indeed because of the cost of the problems that could be caused by failure (e.g. in the banking process, reliability of the transport network, etc.). It should be noted that for several years, account has been taken of the environmental impacts (e.g. with accidental spills of chemical products into the environment, impact on ecosystems, recycling, etc.).

Since the very beginning of research into such systems, the problems linked to validation of those systems have been at the heart of designers' concerns: it is useful to prove the mechanisms to react to the occurrence of failures are well designed, to check that design (by means of simulations, tests, evidence, etc.) and convincingly estimate projected, meaningful values measuring the performances of the functional safety devices.

The difficulty then lies in accurately identifying the various actors involved in the process (users, operators, managers, maintenance personnel, service providers, assessors, authorities, etc.), the different elements in the system, the interactions between those elements, the interactions with the users and the factors which have an impact on the operational safety, ultimately with identification of the electronic and/or programmable elements.

The aim of this first chapter is to offer an examination of the software in the context in which it is used, which is a system, and recap on the links and the constraints which need to be taken into account in creating software.

1.2. Command/control system

Figure 1.1[1] shows an example of a railway system. The Operation Control Center (OCC – photo *a*) controls the whole of the line and passes operational commands to the trains and to the signaling management system (photo *c* shows a manual operation control center).

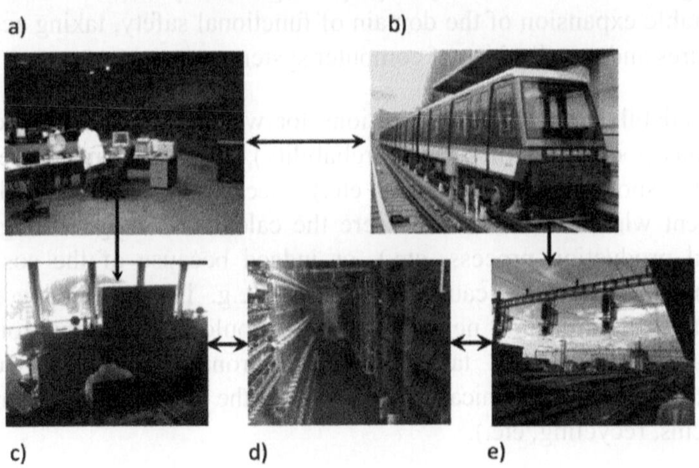

Figure 1.1. *The system in its environment*[2]

1 The picture shows an old-generation operating control center (OCC); new OCCs are stored in PCs and have developed from a physical technology (TCO – optical control view) to display by a video projector.

2 Photos taken by Jean-Louis Boulanger.

The operation control center[3] sends commands to the ground via a set of relays (photo *d* shows an example of a room containing the relays linked to the signaling system). In response to the commands, the ground equipment adopts the desired behavior (in photo *e*, we can see maneuver signals).

Figure 1.1 demonstrates the complexity associated with the concrete system, and highlights the point that a complex system is based not on one piece of software, but on many. Each of these software programs is associated with safety objectives which likely differ from one program to another.

The software involved in supervision does not have as much impact on people's safety as does the software relating to automated control of the trains. For this reason, in the context of systems requiring certification (aeronautics, railways, the nuclear sector, systems based on programmable electronics, etc.), we assign a given level of safety to each software application.[4]

This level of safety is associated with a scale, ranging from "non-critical" to "highly critical". The concept of safety assurance levels and the scales associated therein will be presented in Chapters 2 and 3.

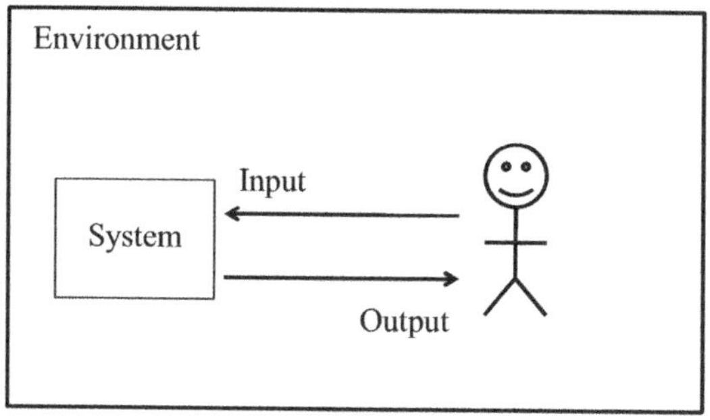

Figure 1.2. *The system in its environment*

Figure 1.2 highlights the fact that the system being constructed is closely linked with an environment which responds to the commands issued by the

3 Figure 1.1 shows a manual operating control center. However, these have now become computerized, and are referred to as PMIs ([BOU 10a – Chapter 5]); PIPCs and PAINGs ([BOU 10a – Chapter 4]).

4 For instance, in the field of aeronautics, the level of safety is called the *Design Assurance Level*; in railways, we speak of the Safety Integrity Level (SIL); and in the automobile sector we have the Automotive Safety Integrity Level (ASIL).

system. It is therefore necessary to acquire a view of the state of the process to be controlled and to have a means of command which is capable of relaying the commands to the environment. The environment may be composed of physical elements, but as a general rule, there are interactions with human parties (operators, users, maintenance personnel, etc.).

During the requirements analysis phase, it is essential to clearly identify all the actors (operators, maintenance personnel, customers, etc.) and identify all the devices which interact with the system. The requirements analysis phase is essential, but can still give rise to numerous omissions and misunderstandings.

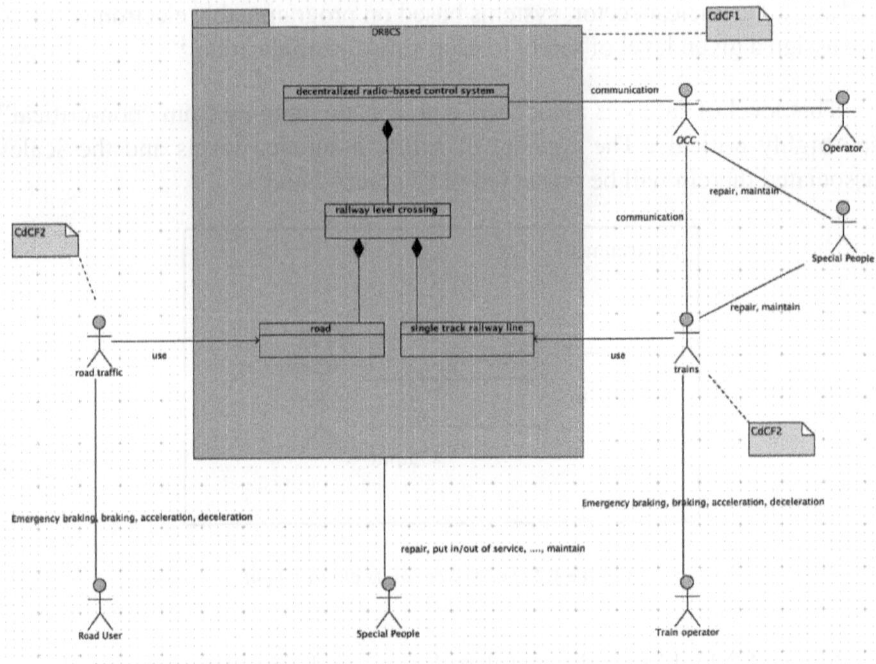

Figure 1.3. *Example of modeling of the system in its environment*

Figure 1.3 presents an example of the modeling of a system to control a level crossing. This system can control the intersection of at least one road with a railway track. This system interacts with various actors (both human and machine): an OCC (as shown in the Figure 1.1), the road users (trucks, cars, etc.), railway users and operators in charge of operation and/or maintenance.

We have chosen to construct a class diagram which models the fact that the decentralized level-crossing management system using a communication system (DRBCS) comprises a level crossing which is itself made up of a railway and a roadway.

The important point in Figure 1.3 lies in identifying the actors which interact with the management system, including the road users, the trains, the OCC and especially the maintenance operators or other personnel (whom the model identifies as "special people").

It is crucial to identify all the actors involved at system level; otherwise there is a risk of forgetting actions – e.g. maintenance activities – but it is also possible to overlook disturbances or malfunctions. We can point to the classic example[5] of the efficiency of a Wi-Fi network, which may correlate to the density of auxiliary networks connected to the system.

Hereinafter, we shall not discuss how to deal with the human factor, because whilst the human factor is an essential one, it does not directly relate to the critical software-based equipment, except for:

– the activities of creation of the software application – hence the need to formalize the skills and responsibilities of the people in charge of the software, as indicated in Chapter 5 of the standard;

– the activities of maintenance and rollout, which are dealt with by Chapter 9 of CENELEC 50128:2011 [CEN 11a].

As regards the identification of the actors involved, it is more usual to speak of identification of the stakeholders; for further information, see Chapter 11 of [BOU 14c].

5 The use of so-called "open" networks (see the standards [CEN 01a] and [CEN 11a]) such as Wi-Fi is attended by a certain number of difficulties, such as network densification (the number of private networks is constantly increasing) and/or interference caused by nearby equipment. It should be noted that, for a very long time, the issue of open networks has not been approached from the standpoint of functional safety, because it relates to aspects such as confidentiality, intrusion, etc., which are covered by the term "security".

1.3. System

Our aim in this section is to lay down the vocabulary relating to the creation of a software-based device. To begin with, we must remember that a software application is directly linked to a device, and that without hardware architecture, there can be no software. Indeed, the validation of a program (see Chapter 5) requires the hardware architecture, and the results are applicable only to that particular hardware. For this reason, the first definitions we shall give relate to the concept of a system and of a software-based system.

DEFINITION 1.1 (System).– *A system is a set of elements interacting with one another, which is organized in such a way as to achieve one or more predetermined results.*

The "organized" part of Definition 1.1 can be seen in the system's organization into different levels, as illustrated by Figure 1.4.

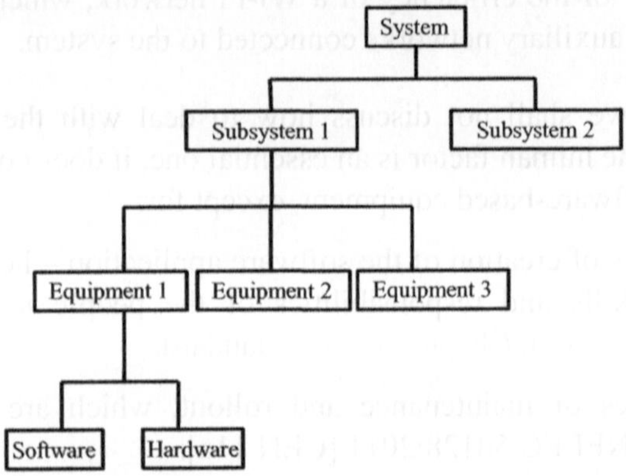

Figure 1.4. *From the system to the software*

Figure 1.4 offers a hierarchical view of the system. This is the view which is used in the railway domain. Hence, a railway line is viewed as a system, which is divided into a number of subsystems: the signaling control subsystem, the passenger transfer subsystem, etc. The signaling control subsystem, for its part, is divided into a number of devices, or classes of equipment: onboard equipment, ground equipment and line equipment.

A system performs several functions. A system function can be subdivided into a variety of subsystems, with each subsystem performing functions which are subfunctions of the whole system's functions. At system level, this representation needs to be accompanied by models which illustrate the interactions between the functions, as shown by the example given in Figure 1.5.

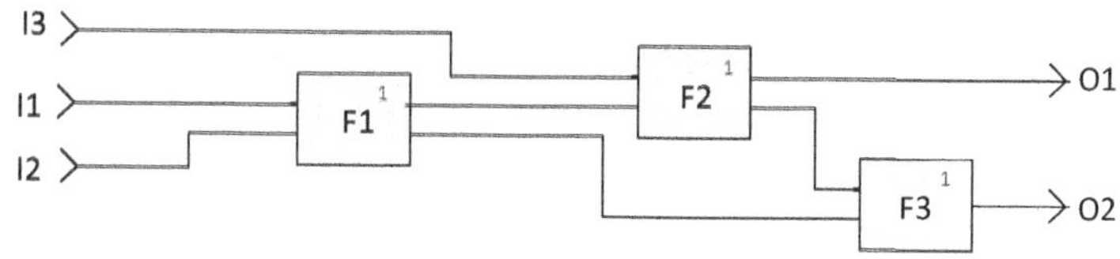

Figure 1.5. *Example of the subdivision of a system*

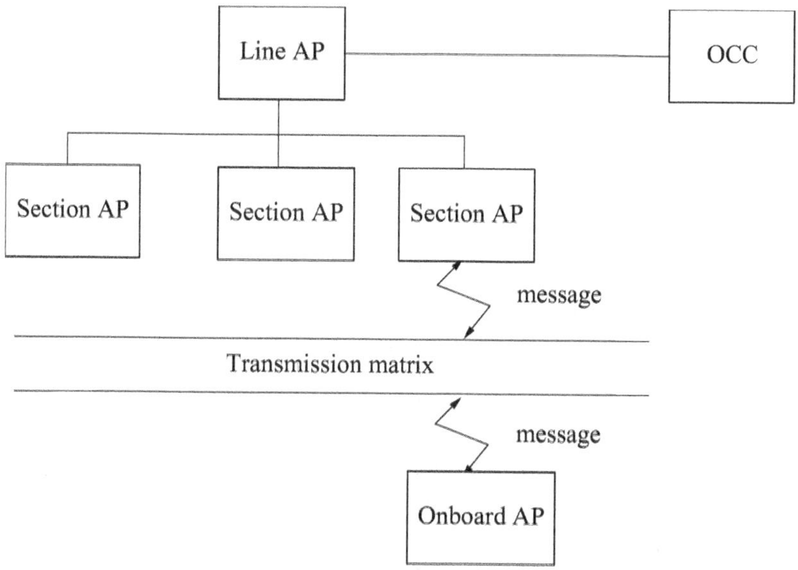

Figure 1.6. *Example of distribution*[6]

6 The example of a subsystem presented is the control system "SAET" used on the "METEOR" line (the rapid-transit East/West Line 14 ("METEOR" is a backronym for this) on the Paris Metro). For further information, see Chapter 2 of [BOU 12].

Thus, a subsystem hosts a variety of functions, which can then be divided between several different pieces of equipment. A piece of equipment is not a functional element in itself; it must be joined by other equipment in order to perform a subsystem-level function.

In terms of a railway system, the difficulty lies in the fact that a train is home to many system functions, and therefore that it contains equipment which contributes to these different functions. For example, the installation of an auto-pilot subsystem involves installing devices on the ground, which communicate with an onboard component on the train, as shown by Figure 1.6.

DEFINITION 1.2 (Software-based system).– *Elements of the system may be totally or partially software-based.*

Figure 1.7 shows that a system is a structured entity (comprising computer systems, processes and use contexts) which must form an organized, coherent whole. Hereinafter, we shall examine the software applications which are found in the computer/automated system component.

In this chapter, we have shown that a system based on programmable electronic equipment is a complex object, which needs to be carefully analyzed in each of its component parts.

1.4. Software application

1.4.1. *What is software?*

In the context of this chapter, the so-called "software" element is a set of computation/processing elements which are executed on a physical hardware architecture so that the system, as a whole, can render the services associated with a device (see Figure 1.4).

Later on in this book, we shall look at the software aspects, so it is necessary, at this point, to define exactly what software is – see Definition 1.3. This definition is slightly different from the one given by ISO 90003:2004 [ISO 04a].

DEFINITION 1.3 (Software).– *Set of programs, processes and rules, and possibly documentation as well, relating to the performance of a set of operations on the data.*

Definition 1.3 does not differentiate between the means (the methods, processes, tools, etc.) used to create the software application, the products created by its execution (documents, analytical results, models, sources, test scenarios, test results, specific tools, etc.) and the software application itself.

This definition is generally associated with the concept of a software application. The concept of software itself is associated with that of executable files.

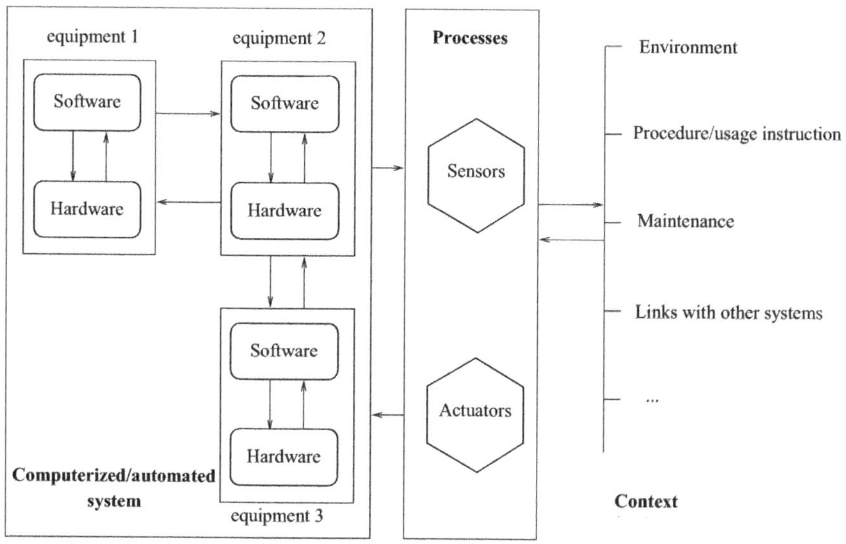

Figure 1.7. *System and interaction*

1.4.2. *Different types of software*

Definition 1.3 shows what the concept of software involves, but it should be noted that there are a variety of different types of software:

– operational software: this term refers to any software delivered to an external customer as part of a program or a product. Test arrays for external usage fall into that category;

– demo: a demonstrator (demo) is a piece of software used by an external customer to help refine their expression of their needs and measure the level of service which could potentially be delivered. These programs are not intended for operational use;

– development tool: a development tool is an internal software application, which is not delivered to an external customer, designed to help development in the broadest sense (editor, compilation chain, etc.), including at the test stage and integration stage;

– model: a model is an internal program for study, not delivered to any external parties, which serves to check a theory, an algorithm or the feasibility of a technique (e.g. by simulation), without the objective of a result or of completeness.

1.4.3. *The software application in its proper context*

In spite of the long-standing monolithic view, we feel it is important to look at a software application as a set of components (see Definition 1.4), which interact to process a set of data. Thus, a component may be a part of the software application, a reused part, a library, a commercial off-the-shelf (COTS[7] – see Definition 1.5) component, etc.

DEFINITION 1.4 (Component).– *A component is an element of software which performs a set of predefined services; these services (or tasks) conform to a clear set of requirements; a component has clear interfaces and is managed in configuration as a separate element in its own right.*

DEFINITION 1.5 (Commercial off-the-shelf – COTS).– *A software product which is available to buy and use without carrying out development activities.*

As Figure 1.8 shows, a software application generally uses an abstraction of the hardware architecture and of its operating system by way of a base layer known as the "base software". In principle, the base software should be written in low-level programming languages such as an assembly language and/or C [KER 88]. It is used to encapsulate the services of the

7 COTS are products that are commercially available and can be bought "as is" (without specification, V&V elements, etc.).

operating system and its utilities, but it also provides relatively direct access to hardware resources.

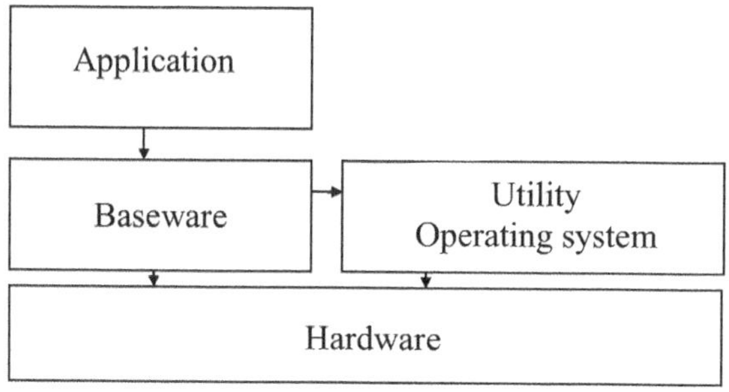

Figure 1.8. *A software application in its environment*

If the software application is associated with a high level of safety, then the base layers (baseware, utilities and operating system) are also associated with safety objectives. The safety objective of the lower layers will depend on the hardware architecture in place (monoprocessor, 2oo2, nOOm, etc.) and on the safety measures employed. In [BOU 10a], we presented real-world examples of safe functional architectures. For the rest of this book, we shall assume that the safety analyses have been performed and that for all the software applications (including the base layer), a level of safety has been allocated.

1.5. Conclusion

In this chapter, we have shown the underlying complexity of programmable electronics-based systems. We have demonstrated the existence of a hierarchy which ranges from the system to the software, through the subsystems, devices and the electronic hardware.

In Chapter 3, we shall show how to control the dependability of a system by controlling the electronic part (see [BOU 10a]) and the software part.

2

Railway Standards

2.1. Introduction

Critical systems for safe operation are characterized by the fact that a fault may have serious consequences, both in terms of people's safety and of economic and/or environmental impact. In many domains, including railways, there are normative frameworks which stipulate that the safety of a newly-created system must be demonstrated. These normative frameworks are supplemented by European and/or national laws and decrees, setting out rules which need to be respected.

These referential frameworks advocate a separation of roles and responsibilities. One team might be in charge of creating the system (development, checking and validation), whilst another would be in charge of demonstrating the safety of that system (conduction of safety studies, drafting of the safety file and analysis of the completeness of the work). A third team, independent of the other two, would be in charge of evaluating the level of safety that is actually delivered. The evaluation may or may not yield a safety certificate.

This chapter is based on the author's experience as a certifying evaluator in the area of railway systems (urban, freight and main lines). In [BOU 07b], we provided an assessment of the implementation of railway standards.

Finally, we shall present a few normative contexts. The aim of this chapter is not to present all of the existing norms and standards and the links between them, but rather to lay down a few points of reference which will be useful for the coming discussion. [BAU 10] offer a presentation of the applicable standards arranged by their domain of application (aeronautics, the automobile industry, railways, the nuclear sector, space and automaton-based systems) and a comparison of those standards (points of similarity and of difference).

2.2. Generic standards

2.2.1. *Introduction*

Electrical/electronic systems have been used for many years, to perform functions relating to safety in the majority of industrial sectors. The IEC 61508 standard [IEC 08] presents a generic approach to all activities linked to the safety lifecycle of electrical/electronic/programmable electronics (E/E/PE) systems which are used to perform safety functions. This standard was published by the International Electrotechnical Commission[1] (IEC), which is an international standardization organization in charge of the domains of electricity, electronics and the related techniques.

IEC 61508 [IEC 08] sets out a generalized approach to safety[2], which is comparable to the ISO[3] 9000 system [ISO 08] on quality.

The IEC 61508 standard [IEC 08] is made up of seven parts:

– IEC 61508-1: general requirements;

– IEC 61508-2: requirements for electrical/electronic/programmable electronic safety-related systems;

– IEC 61508-3: software requirements;

– IEC 61508-4: definitions and abbreviations;

– IEC 61508-5: examples of methods for the determination of safety integrity levels;

1 For further information, see www.iec.ch/.

2 IEC 61508 [IEC 08] does not cover the aspects of confidentiality and/or integrity (i.e. *security*), which relate to the establishment of precautions to prevent unauthorized persons from harming and/or affecting the safety provided by the E/E/PE safety-related systems. In particular, such precautions involve monitoring of the networks in order to prevent intrusion.

3 See www.iso.org/iso/home.html.

– IEC 61508-6: guidelines on the application of IEC 61508-2 and IEC 61508-3;

– IEC 61508-7: overview of techniques and measures.

The IEC 61508 standard [IEC 08] is absolutely coherent with the convergence which is observed between the various industrial sectors (aeronautics, nuclear, railway, automobile, manufacturing, etc.), but the content of IEC 61508 is so complex (or unusual) that it is necessary to be guided through the standard. Thus, interested readers should refer to [ISA 05] or [SMI 07].

In most cases, safety is accomplished by many systems of diverse technologies (mechanical, hydraulic, pneumatic, electrical, electronic, programmable electronics). The safety strategy must take account of all the factors which contribute to safety. Thus, IEC 61508 [IEC 08] sets out an analytical framework which applies to safety-related systems based on other forms of technology, before going on to specifically discuss electronics-based systems.

Owing to the wide variety of E/E/PE applications and their greatly-varying degrees of complexity, the exact nature of the safety measures to be taken depends on factors specific to the application; for this reason, in IEC 61508 [IEC 08], there are no general rules, but rather recommendations regarding the analytical methods that should be used.

2.2.2. *Safety levels*

IEC 61508 [IEC 08] defines the concept of the Safety Integrity Level (SIL). It should be noted that a similar concept exists in all domains, although the vocabulary used varies from one field to another. In aeronautics, one speaks of the Design Assurance Level (DAL) and in other domains, of the SIL. In the automotive industry, the concept of the Automotive SIL (ASIL) is used.

The SIL offers a scale against which to measure and quantify the safety level of a system. Safety Integrity Level 1 (SIL 1) is the lowest level possible, and SIL 4 the highest. The four SILs for a safety function are characterized by the potential impact of a failure:

– SIL 4: catastrophic impact (highest level);

– SIL 3: impact on the community;

– SIL 2: major protection of the installation and of production, or risk of injury to employees;

– SIL 1: minor protection of the installation and of production (lowest level).

The standard details the requirements which must be fulfilled in order to reach each safety integrity level. These requirements are more demanding for higher SILs, with a view to ensuring a lesser probability of a dangerous fault.

Using the SIL scale, it is possible to set out prescribed practices relating to the integrity of the safety functions being entrusted to the E/E/PE systems.

2.3. History between CENELEC and the IEC

Electric/electronics systems have been in use for many years, performing safety-related functions (the term *safety* is used in this context in most industrial sectors). IEC 61508 presents a generic approach to all activities pertaining to the safety lifecycle of the E/E/PE systems used to perform safety-related functions.

Figure 2.1 shows a link between the IEC 61508 [IEC 08] and IEC 61513 [IEC 11] standards, but this link is inaccurate. In actual fact, standards were in place in the nuclear industry long before IEC 61508 was written, and the link exists only because of the naming system.

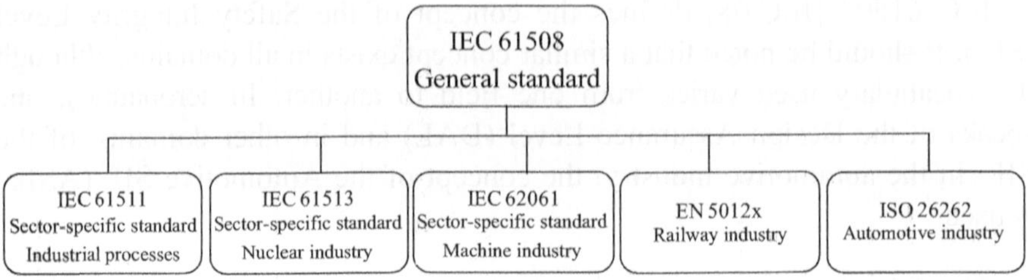

Figure 2.1. *IEC 61508 and its offshoots*

2.4. CENELEC referential framework

Today, railway projects are governed by texts (laws, decrees, declarations, etc.) and a normative framework (CENELEC[4] EN 50126 [CEN 99], EN 50129 [CEN 03] and EN 50128 [CEN 01a, CEN 11a]) which intend to define and achieve certain objectives in terms of RAMS (Reliability, Availability, Maintainability and Safety).

The three standards cover the safety-related aspects of the system, down to the hardware and/or software elements used. Figure 2.2 illustrates the relationships between the various CENELEC standards.

2.4.1. *Introduction*

As Figure 2.1 illustrates, the railway standard CENELEC EN 5012x is an offshoot of the generic IEC 61508 safety standard [IEC 08] which integrates the specific features of the railway domain and of the successful experiments (SACEM[5], TVM[6], SAET-METEOR [MAT 98], etc.).

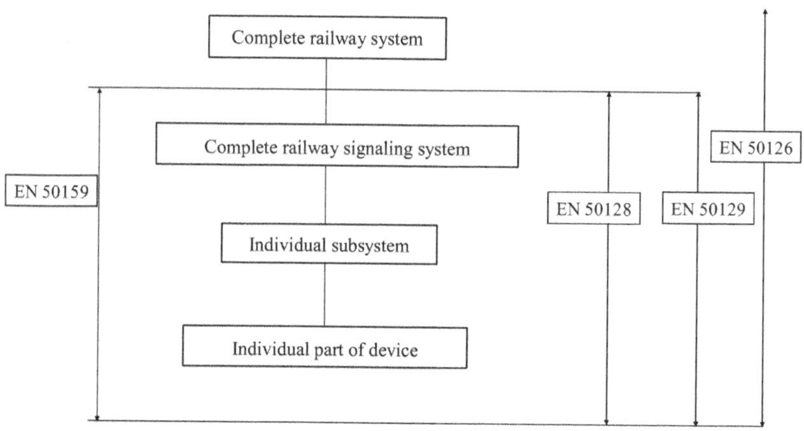

Figure 2.2. *Main standards applicable to railway systems*

4 *Comité européen de normalisation électrotechnique* – see www.cenelec.eu/.

5 SACEM, which stands for *Système d'Aide à la Conduite, à l'Exploitation et à la Maintenance* (Driving, Operation and Maintenance Support System), is a system used on Line A of the Paris RER (suburban rapid rail) to display signaling data in the driver's cabin.

6 TVM, which stands for *transmission voie-machine* (Track-to-Machine Transmission), is a railway signaling system in use in the driver's cabins on France's high-speed train lines.

The railway standard CENELEC EN 5012x is applicable both to urban rail applications (e.g. the Paris metro and RER, etc.) and to conventional rail applications (high-speed lines, ordinary trains, freight trains).

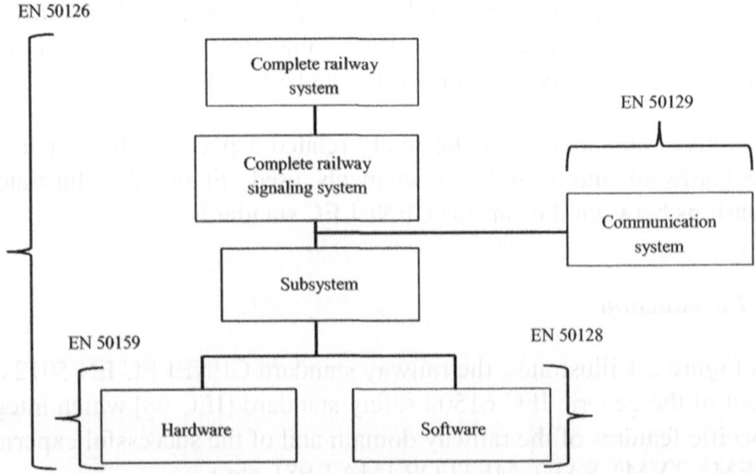

Figure 2.3. *Organization of standards applicable to railway systems*

One of the restrictions on this railway standard (CENELEC EN 5012x) relates to the domain of application of CENELEC EN 50128 [CEN 01a, CEN 11a] and CENELEC EN 50129 [CEN 03], which are normally limited to the signaling subsystems (see Figure 2.3). For the hardware architectures used in other subsystems, there are previous standards which remain applicable (NF F 00-101, etc.). Work is currently under way at the various certification bodies to extend and generalize these two standards to the whole of the railway system.

2.4.2. *Description*

In the area of railways, the normative referential framework comprises the following standards:

– CENELEC EN 50126 [CEN 99], which describes the methods to be used to specify and demonstrate reliability, availability, maintainability and safety (RAMS);

– CENELEC EN 50128 [CEN 01a, CEN 11a], which describes the actions to be taken in order to demonstrate the safety of the software;

– CENELEC EN 50129 [CEN 03], which describes the structure of the safety file and the means needing to be mobilized to ensure the safety of the hardware.

Thus, CENELEC EN 50129 is oriented more specifically toward the management of the safety of the hardware architectures. In [BOU 10a], we presented the principles of safety management for hardware architectures, and we presented examples of safe, functioning architectures.

CENELEC EN 50128 [CEN 01a, CEN 11a] pertains to the management of the safety of the software applications. In [BOU 11b], we presented safety management techniques for a software application.

These standards are complemented by a standard relating to the "transmission" aspects. The 2001 version of CENELEC EN 50159 was divided into two parts: EN 50159-1 [CEN 01b], devoted to closed networks, and EN 50159-2 [CEN 01c], devoted to open networks. The new version of CENELEC EN 50159 [CEN 11b] supersedes the previous version, and covers all types of networks (both open and closed).

Thus, CENELEC EN 50159 also covers some of the aspects relating to *security*. It has emerged, in the wake of the advent of new uses (e.g. the use of Wi-Fi networks) that this standard is no longer sufficient, because it does not relate to the management of COTS or to physical defenses against intrusion into premises.

It is worth pointing out that in addition to these standards pertaining to the architecture of a device (software + hardware), there are further standards which regulate the device's interactions with the environment, such as CENELEC EN 50121 [CEN 06], which deals with the issues of electromagnetic compatibility (EMC), both from the point of view of EM radiation emitted from the railway system into the outside world and from the point of view from the environment into the railway system (affecting the train and fixed equipment).

Figure 2.4 examines the safety lifecycle in the context of the railway domain. The cycle is divided into three stages:

– preliminary analyses (stages 1, 2 and 3);

– creation of the system, the subsystem and/or equipment (stages 4–9);

– commissioning and use of the system (stages 10–14).

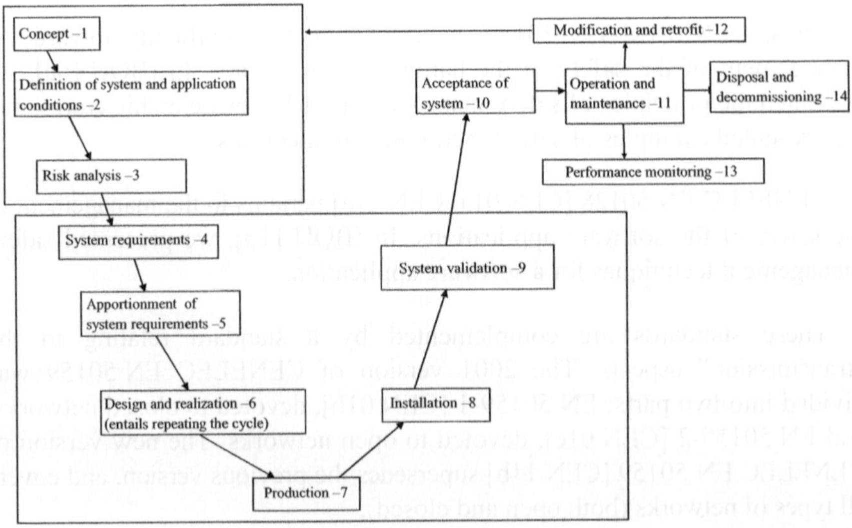

Figure 2.4. *Safety lifecycle*

In CENELEC EN 50129 [CEN 03], system requirements are divided into safety-related and non-safety-related requirements. CENELEC EN 50126 [CEN 99] suggests the following steps (here we are only looking at the aspects of prototyping/production of the system) to outline and demonstrate the safety of a system:

– define the system and application conditions: draw up a mission profile, describe the system, create an operation/maintenance strategy and identify any constraints caused by the pre-existing infrastructure (other systems or other lines developed previously);

– perform a risk analysis;

– in terms of the system requirements: analyze the requirements, detail the system and its environment, and define the system's demonstration and acceptance criteria;

– allocation of system requirements: specify the subsystem requirements, the equipment and/or component requirements, and define the acceptance criteria for these elements;

– design and implementation: carry out the design and development processes, and perform the necessary checks and validation.

2.4.3. *Implementation*

The purpose of the CENELEC framework is to:

– provide a Europe-wide common framework in order to encourage the expansion of the railway-equipment market and enhance the interoperability, interchangeability and cross-acceptance [CEN 07] of railway equipment;

– deal with the specific requirements of the railway domain.

Safety, in the sense of operational safety, is assured by the use of predefined concepts, methods and tools, throughout a system's entire lifecycle. A safety study entails an examination of the system's faults. We must identify and quantify the seriousness of the potential consequences and, as far as possible, the frequency with which such faults can be expected to occur.

Amongst the risk-reduction measures which can be taken to ensure the level of risk is acceptable, CENELEC EN 50129 [CEN 03] describes the setting of safety objectives (see Chapter 3) for the functions of the system and of its subsystems. The Safety Integrity Level (SIL) is defined as one of the discrete levels to specify the safety integrity requirements for the functions assigned to the safety-related systems.

More specifically, CENELEC EN 50128 [CEN 01a, CEN 11a] is devoted to the aspects of software development for use in railway systems. In reference to the software, the Software SIL (SSIL) defines varying levels of criticality, ranging from 0 (no danger) to 4 (critical).

Remember that a definition of what is meant by "software" (Definition 1.3) was given in section 1.4.1.

Safety requirements have always been taken into account in complex systems (rail transport, air transport, nuclear power plants, etc.). Today, though, contractual obligations as regards performances have led industrialists in the railway sector, for example, to very carefully define the parameters affecting a system's reliability, availability and maintainability (RAM). The choice of norms and standards is the responsibility of the designer and/or manufacturer.

2.4.4. *Software safety*

CENELEC EN 50128 [CEN 01a, CEN 11a] specifies the procedures and technical prerequisites applicable to the development of programmable electronic systems used in railway control and protection applications. CENELEC EN 50128 is therefore not normally applicable to all software applications in the rail sector.

EN 50155 [AFN 07], on the other hand, is normally applicable for all onboard applications on a train.

2.4.5. *Safety versus availability*

The standards in force describe the processes, methods and tools necessary to attain and demonstrate the required SIL. It is a question of obligatory techniques, which applies in addition to the obligations in force regarding the quantitative and/or qualitative results.

At this early stage, we have chosen to focus on safety, because safety is the main aspect of operational fitness which is crucial to analyze. There are few, if any, requirements in terms of reliability or availability which are set out by the normative framework.

In the context of critical systems (SIL 3 and 4), the principles of safety implementation run counter to the principles of system availability. For example, in the area of railways, the safety status which is "train stopped" has a significant impact on the system's overall availability. On the other hand, the obligations governing the creation of a critical system help to improve the reliability of the software. This reliability is achieved by conditioning the complexity and quality of the software.

However, in the context of non-critical (SIL 0) and low-criticality software applications (SIL 1 and SIL 2), the process of software development is subject to less stringent constraints (in terms of the choice of programming language, the tools and the processes used). This lessens the quality of the software, leading to software packages whose reliability is poor. More specifically, in the context of SIL 0, 1 or 2 systems, the industrialists can use "Commercial Off-The-Shelf" (COTS) products. Quality assurance, which has a direct impact on the reliability and availability of COTS products, remains a current issue.

2.5. EN 50155 standard

EN 50155 [AFN 07] is applicable for all electronic equipment[7] fitted in railway vehicles. This standard is not solely devoted to software applications: section 7.3 therein gives an in-depth description of the rules applicable to software-related aspects.

The first requirement relates to the obligatory conformance to ISO 9001:2008 [ISO 08]. The second relates to the putting in place of configuration-management procedures, which must cover the software applications developed but also the tools used for development and maintenance.

The third requirement relates to the definition of a lifecycle for the development of a software application, which must be structured and documented.

As illustrated by Figure 2.5, the development cycle advocated by EN 50155 [AFN 07] is divided into five stages:

- specification of the software;
- design of the software;
- testing of the software;
- testing of the hardware/software integration;
- maintenance of the software application.

7 In this context, "electronic equipment" includes devices used for command, regulation, protection, power supply, etc.

For the specification phase, EN 50155 [AFN 07] introduces the concept of "requirements" and that of a "specifications list", but does not go into any greater depth about the concept of requirements.

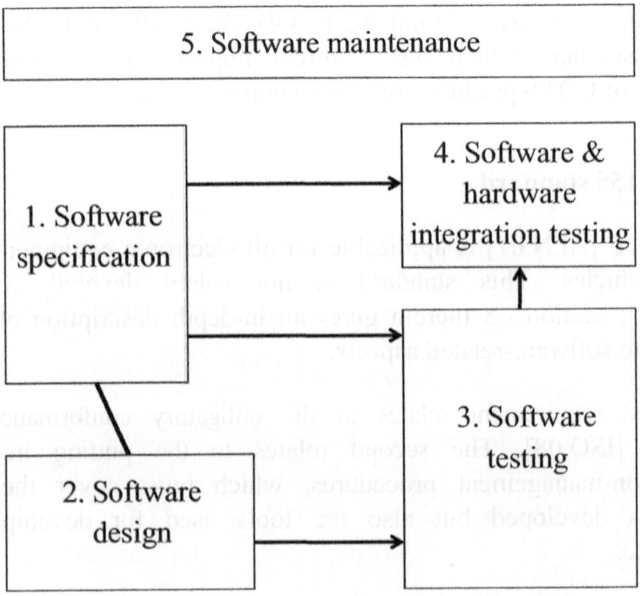

Figure 2.5. *Development cycle advocated by EN 50155*

As regards the software-development phase, section 7.3.2 of EN 50155 [AFN 07] states that a modular approach is needed (i.e. the software is divided into small components – or "modules" – which limits the size and the number of interfaces). The standard also prescribes a clearly-defined approach (procedure, list of documentation needing to be produced, type plan, standardization, etc.) and a structured method (logic/function block diagrams, sequence diagrams, data flow diagrams, truth tables/ decision tables). With regard to the code, the standard states that it is necessary to choose a programming language which offers easy verification (requiring a minimum level of effort for understanding). The development method and the programming language should make it easy to analyze the code. The program's behavior should be able to be deduced from analyses of the code.

With regard to testing, EN 50155 [AFN 07] recommends using the technique of boundary-value analysis, equivalence class partitioning and input domain partitioning. Test cases may be selected using a model and a simulation of the process.

The maintenance phase is a key point: it is essential to guarantee that any modification, addition of a function or adaptation of the software application – e.g. for a different type of train – does not compromise safety. The management of the maintenance phase must be fully defined and documented. One of the difficulties with the maintenance phase is to manage the simultaneous evolution of multiple parallel versions of the software, and installation on the ground and aboard the fleet (as the trains are of different types and are parked in different places).

EN 50155 [AFN 07] recommends that all the elements produced be recorded in a format that facilitates subsequent analysis. With regard to the documents that must be produced, the standard gives the following list:

– software requirements specification with a demonstration that the system requirements are fulfilled;

– a description of the architecture and the design of the software application capable of serving the requirements expressed in the software requirements specification;

– for each software component, it is necessary to give a description of the execution characteristics (list of inputs and outputs, etc.), the source code, the description and the test results;

– the dataset for the software application (global variables and global constants) must be described in a document known as the "data dictionary";

– a description of the memory embedding of the software application;

– a description and a reference (name, version) of the tools used to develop the software application and produce the executable file;

– a description of the integration tests and of the associated results.

In sections 7.3.2 and 10.2, the EN 50155 [AFN 07] standard states that the CENELEC 50126 [CEN 99] and 50128 standards [CEN 01a, CEN 11a] are applicable in this area, but does not clearly define in what way they are applicable, or the precise extent of that applicability.

2.6. CENELEC 50128

2.6.1. *Introduction*

CENELEC 50128 [CEN 01a, CEN 11a] applies for both safety-related and non-safety-related domains – for this reason, CENELEC 50128 introduces SSIL 0, which pertains to non-safety-related software applications – and applies exclusively to software and the interaction of a software application with the whole system.

This standard recommends the implementation of a V-lifecycle, from the software specification to the software testing. One of the peculiar points of CENELEC 50128 lies in the need to deconstruct the implementation of the resources. For this reason, it is said to be a “resources standard”.

CENELEC 50128 explicitly introduces the concept of evaluation. As we showed in [BOU 07a], for software applications, evaluation involves demonstrating that the software application achieves the safety objectives set for it.

Evaluation of the SIL of a software application is based on:

– the use of audits: is quality-assurance (QA) effectively employed on the project?

– reviewing of the plans (software QA plan, testing plan, verification and validation (V&V) plan, etc.);

– reviewing of the elements produced (documents, source code, generation chain for the executable, data production process, test scenarios and results, safety assessment (e.g. Software Error Effect Analysis (SEEA)), etc.);

– formulation of comments and potential non-conformities;

– formulation of an assessment in the form of an evaluative report.

2.6.2. *SSIL management*

In Chapter 4, we shall present the approach to safety management as proposed by CENELEC 5012x. In this section, we shall focus on how to identify the SSIL on the basis of the SIL, and outline the general ideas.

2.6.2.1. *SSIL attribution*

The management of software safety involves management of the SSIL. As indicated above, the value of the SSIL ranges between 0 (no impact whatsoever on safety) to 4 (major impact on safety).

Figure 2.6 shows the breakdown of the set Tolerable Hazard Rate (THR) either for a subsystem in terms of Failure Rate (FR), or for the hardware making up that subsystem in accordance with the architecture (in which case, independence hypotheses may be made), and the division of the SIL either for each subsystem in terms of SIL (giving us the SSIL) or for the hardware (or software) components:

– the hardware's SIL is identical to the SIL of the subsystem;

– the software's SSIL is identical to the SIL of the subsystem.

The switch from a subsystem SIL to a device SIL is made with the rules introduced above, but it is possible to take account of the architecture of the device to refine the SILs and SSILs. Thus, the SSIL is linked to an SIL rather than to a degree of probability.

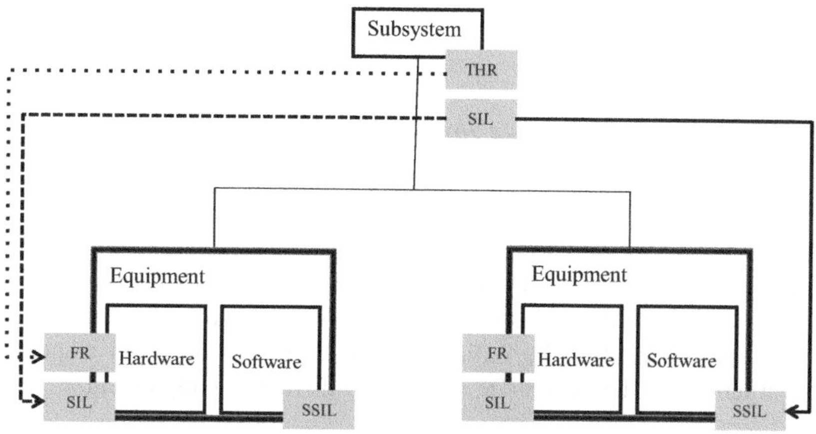

Figure 2.6. *Switch from the THR/SIL for the subsystem to the SSIL for the software package*

2.6.2.2. *Choice of SSIL*

The SSIL is set on the basis of safety analyses as an objective to be attained in the creation of the software. Thus, CENELEC 50128 introduces the concept of independent evaluation to confirm that this objective has been achieved.

The SSIL is a safety integrity objective which must be met. This objective cannot be quantified, which is why CENELEC 50128 includes the idea of evaluation. An outsider, independent of the software development team, evaluates whether or not the objective has been achieved. It must be remembered that the main tactic in creating safety-related software is to monitor the number of system faults, and in order to do so, it is necessary to monitor the development process and the complexity of the software. The SSIL is tied in with the concept of confidence, and therefore the question asked of the evaluator is: do you have confidence in the performance of this program?

An in-depth analysis of CENELEC 50128 shows that for SSIL1 and SSIL2, the same type of activities are performed; the same is also true for SSIL3 and SSIL4. Remember that SSIL1 relates not to impacts on human life but rather to impacts on the equipment, which is why three levels of SSIL are generally used: SSIL0, SSIL2 and SSIL3/4. If an SSIL1 is identified, it is recommended that it be reclassified as either an SSIL0 or an SSIL2 depending on the seriousness of the risks associated therewith.

2.6.3. *Comparison of 2001 and 2011 versions*

2.6.3.1. *CENELEC 50128:2001*

The structure of the 2001 version of CENELEC 50128 [CEN 01a] is illustrated in Figure 2.7. It should be noted that in this version of the standard, section 17 describes the management of the software settings. This section was added after the initial draft. Therefore it does not respect the same structure as Chapters 8–16, and is very difficult to implement.

Section 16 introduces the basic concepts of software maintenance, whilst still being orientated toward "bug correction". In this version, it should be noted that there is a low degree of common activities (tests, verifications, etc.).

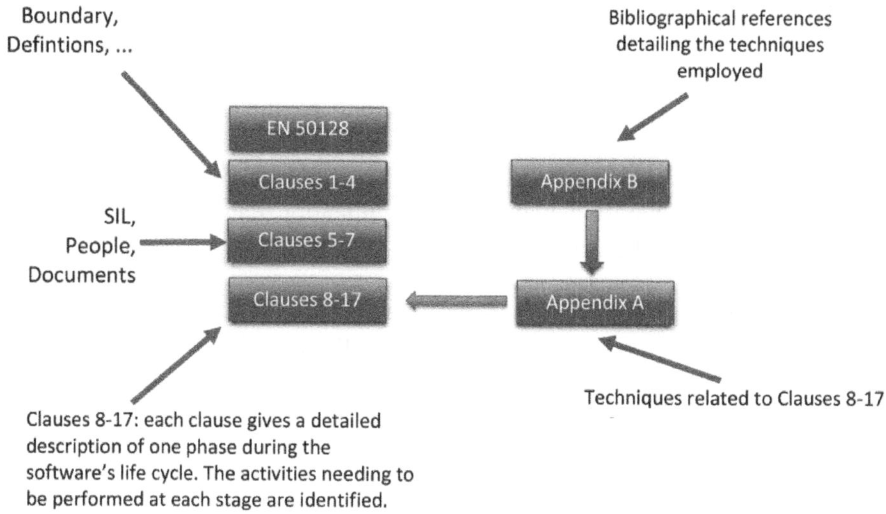

Figure 2.7. *Structure of CENELEC EN 50128:2001*

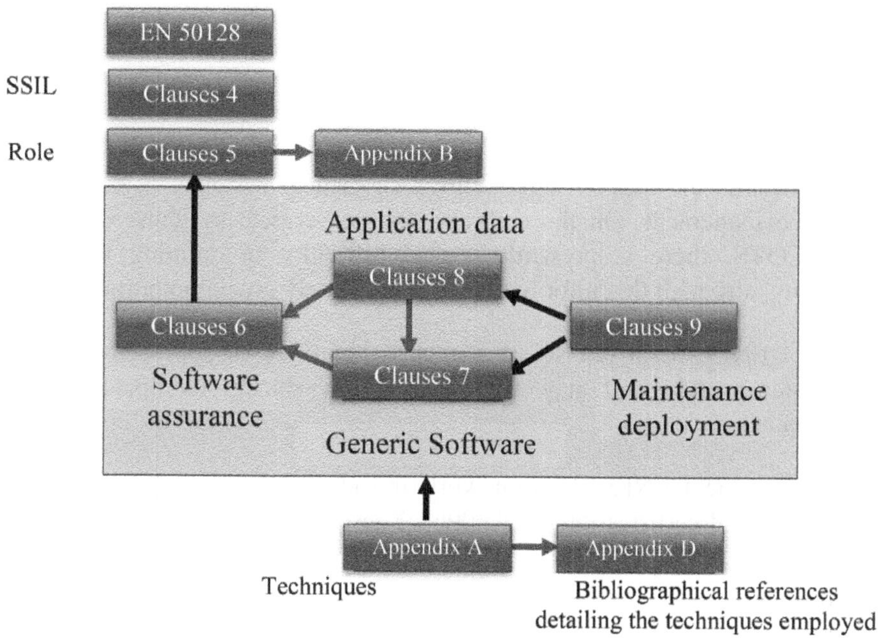

Figure 2.8. *Structure of CENELEC EN 50128:2011*

2.6.3.2. *EN 50128:2011*

The 2011 version of CENELEC 50128 [CEN 11a] introduces the concept of software assurance (SwA), which is aimed at creating a software package with a minimum level of error, based on a Quality Assurance, skill evaluation, verification and validation, and independent evaluation. It introduces clear separation between the application data and the software, which is then called the generic software.

One of the important points in the new version of CENELEC 50128 is the addition of Clause 9, which is devoted to the oversight of the software's maintenance and deployment. Figure 2.8 illustrates the structure of the 2011 version of CENELEC 50128.

2.7. Conclusion

From the above, we can conclude that, irrespective of the specific domain, there is a referential framework in place which gives a scale that can be used to quantify a system's level of criticality. This level of criticality helps define the amount of effort that must be invested. Safety management for a critical system involves defining processes based primarily on testing activities.

As regards the aspect of reliability, for critical systems, it stems from quality management. On the contrary, for non-critical systems, which are often COTS, there is presently a real difficulty in defining their true reliability, although this information is of absolutely crucial importance.

The CENELEC 5012x set of standards identifies an approach to create the system safely and stay abreast of safety during its operation and maintenance.

Finally, the CENELEC framework introduces the concept of product evaluation and certification [BOU 09b]. These concepts will be explored in detail in Chapter 10.

3

Risk and Safety Integrity Level

3.1. Introduction

The aim of this chapter is to present the concepts of risk, undesirable events, accidents, severity, frequency and level of risk. Here, we are able to set out a clearly-defined lexicon, and also point out the differences between the vocabulary used in different domains.

For example, in aeronautics, we speak of the Design Assurance Level (DAL), and in other domains, the concept used is the Safety Integrity Level (SIL). For the automotive industry, the concept of ASIL[1] is used, and there are two new criteria – exposure and controllability – which play a part in the definition of the risk level.

3.2. Basic definitions[2]

In the context of railway systems, we seek to prevent or avoid accidents. An accident (see Definition 3.1) is a situation which may be associated with various types of damage/injury (individual, collective).

DEFINITION 3.1 (Accident – CENELEC 50129).– *An accident is an unexpected event or series of events causing death, injury, loss of a system or service, or damage to the environment.*

---

1 Automotive SIL – the automotive industry has adapted the qualifier levels to take account of the specificities of that domain, for more information see [ISO 11].

2 This section is repeated from [BOU 06].

By its very nature, an accident, in order to be prevented, must be taken into account as an eventuality in the creation of the system. In land transport, as pointed out by [BAR 90], accidents can be classified into a number of different categories:

– system accidents: the users of the system are passive, and are caused harm, by the fault of the staff (maintenance, intervention, operation, etc.), failures within the system or abnormal conditions in the environment;

– user accidents: this category of accidents relates to events which are localized to one or a few users (e.g. illness, panic, suicide, etc.) and which have very little to do with the system;

– system/user accidents: unlike with system accidents, in this case, the user is active and interacts with the system. The accident is caused by an improper use of the system (e.g. failure to heed the audio warning of the doors closing).

From the definition of an accident, we see a new concept appear: that of an *event*. This can be refined somewhat by introducing the concept of an *undesirable event* (see Definition 3.2).

DEFINITION 3.2 (Undesirable event).– *An undesirable event is an event which should not be allowed to occur, or which should have only a very low level of probability of occurring.*

The harm caused can also be classified into two categories: collective harm and individual harm.

A user accident is, by nature, likely to cause individual harm, whereas the other two categories of accidents may cause individual or collective harm. With this understood, it is then possible to define a level of severity[3]: insignificant (minor), marginal (significant), critical and catastrophic.

In [HAD 98], which applies only to rail transport, the authors offer a typology of accidents which links the categories of accidents, the potential accidents, the types of harm, the level of severity, the types of hazard and the hazardous factors. Using this typology, it is possible to show the link

3 We have chosen to focus on the railway domain [CEN 99], but where alternative terms exist and are used in other domains, they are given in parentheses.

between a hazardous factor, such as a lightning strike, and a system accident such as a collision.

Whilst an accident is, by definition, conclusive (i.e. the event must already have occurred in order for the situation to be called an accident), there is an intermediary situation: a *potential* accident. The concept of a potential accident refers to a known situation which could lead to an accident.

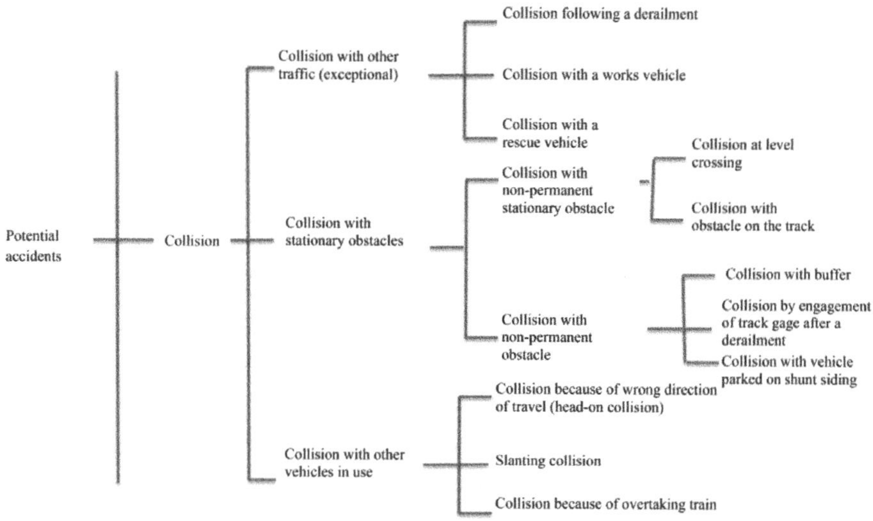

Figure 3.1. *Example of classification of a potential accident*

For railway systems, the list of potential accidents (see Definition 3.3) contains: a derailment, a passenger fall, a collision, dragging of a person, trapping of a person, electrocution, fire, explosion, flooding, etc.

DEFINITION 3.3 (Potential accident).– *A potential accident is an unexpected event or series of events which could lead to an accident in the wake of the occurrence of an additional event for which the system is unprepared.*

Figure 3.1 is reproduced from [HAD 95]. It shows an example of the classification of a potential accident relating to a collision. Table 3.1 gives a preliminary list of potential accidents. The situation of a potential accident is connected to the concept of a hazard (see Definition 3.4).

Potential accident	
Passenger fall	The passenger falls from the train onto the platform The passenger falls, on the train, whilst at a station The passenger falls from the platform onto the track
Passenger injury	The passenger cannot disembark The passenger is caught between the train doors and the platform The passenger is on the track and is hit by a train …
Derailment	Derailment due to excess speed Derailment due to an incorrectly-positioned signal …
Collision	Collision between trains Collision with an obstacle
Staff injury	Staff on the track are hit by a train …
Asphyxia/poisoning	…
Burning	…
Electrocution	…
Fire	Fire on the train Fire outside of the train

Table 3.1. *Example of list of potential accidents*

DEFINITION 3.4 (Hazard – CENELEC 50126).– *A hazard is a condition which could give rise to a potential accident.*

Thus, a hazard (see definition 3.4) is a hazardous situation, the consequences of which could harm people (injury or death of persons), the company (loss of production, financial loss, loss of image, etc.) or the environment (degradation of the natural world, animal habitats, pollution, etc.). Depending on whether the cause is random or deliberate (deterministic), the terms *hazard* and *threat*, respectively, are used. Thus, the term *threat* is used in the context of security activities.

Hazards can be classified into three categories: hazards caused by the system or its components, hazards caused by humans (failure on the part of the operating staff; failure on the part of the maintenance staff; intervention by passengers) and hazards caused by abnormal circumstances in the environment (earthquakes, wind, fog, heat, humidity, etc.).

The identification of hazardous situations for a given system involves carrying out systematic analyses which include two complementary phases:

– an empirical phase, based on feedback from the ground (a pre-existing list, accident analysis, etc.);

– a creative and/or predictive phase, which could be based on brainstorming, projected studies, etc.

The concept of a hazard (a hazardous situation) is directly connected to the concept of failure.

Figure 3.2 represents the path which leads from the undesirable event to the accident. It should be noted that the potential accident can only come to pass in a suitable operational context. Indeed, if the hazard is a driver ignoring a red light on a secondary track with little traffic, the risk incurred is less than it would be on a busy main line. Similarly, the path which leads from an undesirable event to a hazardous situation is linked to a technical context.

Figure 3.3 illustrates the complete chain of events between the cause and the potential accident (for further information, readers could consult [BLA 08], for example). The concept of cause is introduced to form the link with the different types of failures to which the system may fall prey, such as the failure of one or more functions (a function may extend transversally to several internal pieces of equipment), failure of one or more such components, human error and external factors (EMC, etc.).

Figure 3.4 shows how, for a potential accident, factors come together and combine, with a definite cause-and-effect web being woven. It is then the responsibility of the analysts to select the representative (Cause/Undesirable Event/HazardousSituation/PotentialAccident) scenarios.

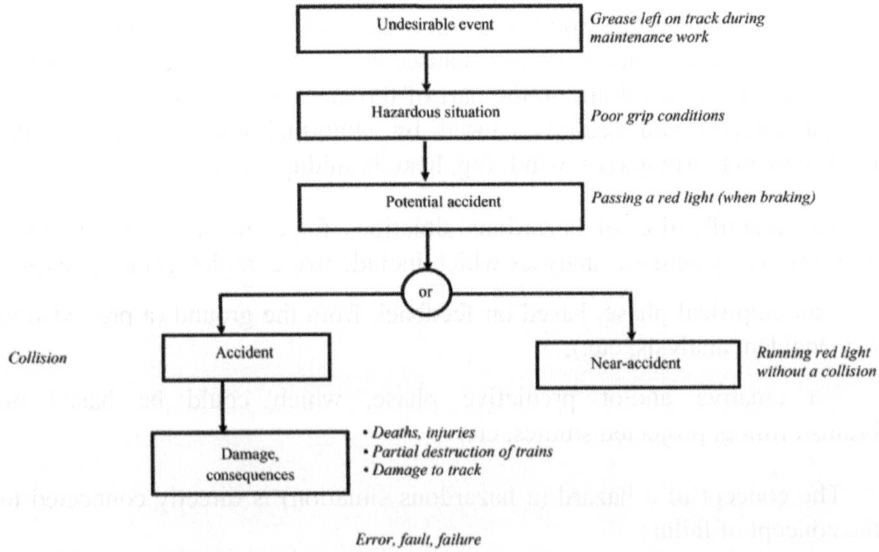

Figure 3.2. *Chain of events leading up to an accident*

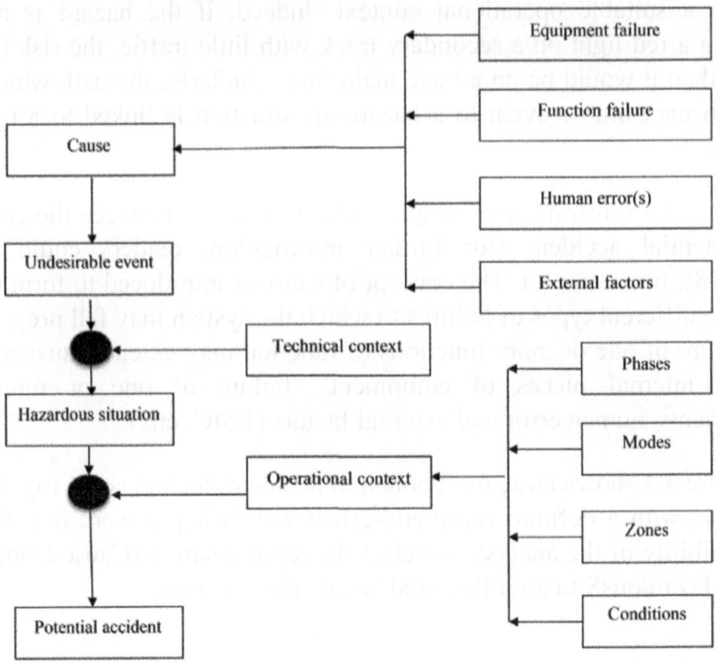

Figure 3.3. *Chain of events leading up to a potential accident*

3.3. Safety enforcement

In this section, we shall discuss and present certain concepts pertaining to safety. For the most part, the perspective adopted will be that of the railway domain, but for certain concepts we shall use definitions drawn from other domains if they are relevant.

3.3.1. *What is safety?*

In the field of rail transport, two safety enforcement principles are used (see [BIE 98]): probabilistic safety (see Definition 3.6) and intrinsic safety (see Definition 3.5).

DEFINITION 3.5 (Intrinsic safety).– *A system is said to be intrinsically safe if we can be certain that any failure of one or more components of that system will only result in its becoming more permissive than it was at the time the failure occurred. Note that in rail transport, a complete stoppage of the train is generally the most restrictive state.*

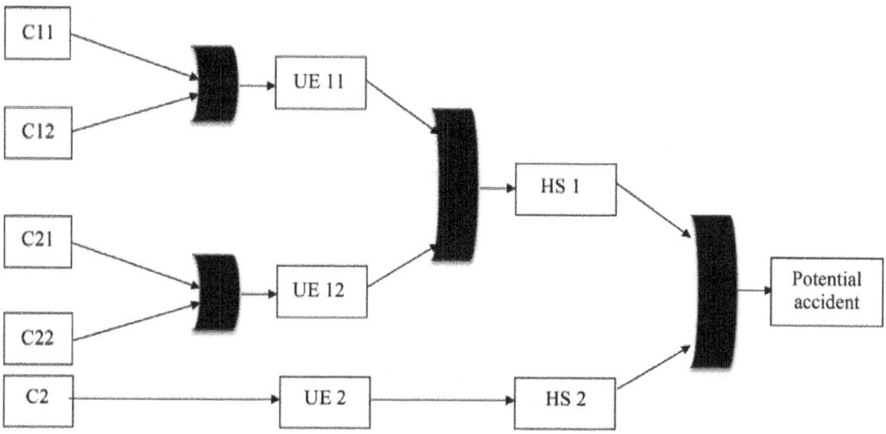

Figure 3.4. *Combination of events leading up to a potential accident*

Intrinsic safety is based on the physical properties of the components. Hence, we need to know exactly how the components will behave in case of one or more faults. Therefore, the components used in intrinsically-safe systems tend not to be overly complex.

DEFINITION 3.6 (Probabilistic safety).– *Probabilistic safety involves being able to demonstrate, a priori, that the probability of occurrence of a hazardous situation is lower than a predefined tolerable threshold.*

Thus, probabilistic safety entails acceptance of the fact that there is no such thing as zero risk. The concept will be explored in greater detail later on. The combination of probabilistic safety and intrinsic safety enables us to define controlled safety. A device is said to have controlled safety if it is capable of detecting a fault which could lead to a state whose consequences would be unacceptably severe, and in this case, enters a safe mode.

DEFINITION 3.7 (Computer security).– *Computer security is defined by the set of means and resources employed to minimize the vulnerability of a computer system to accidental or intentional threats.*

Two very different types of threats can be discerned: accidents and deliberate attacks. A distinction also needs to be drawn between the prevention of physical harm (*Safety*) and the prevention of intrusion (*Security*).

Safety relates to the protection of computer systems from accidents due to the environment, or system errors. Safety is a priority in computer systems which control real-time processes which could endanger human lives (automated transport, nuclear power plants, etc.). Real-time systems are subject to very stringent time restrictions.

DEFINITION 3.8 (Safety).– *Safety relates to the prevention of catastrophes (harm prevention).*

The defense of computer systems against malicious actions (intrusions, vandalism, etc.) falls under the remit of "security". In general, security relates to systems which perform sensitive processes or handle sensitive data (e.g. bank-account management, etc.).

With the introduction of new technologies in computer-based systems, the above is no longer entirely true. In the area of rail transport, technologies such as Wi-Fi or GSM (or sometimes GSM-R) are being introduced, and measures will need to be taken to ensure security.

DEFINITION 3.9 (Security).– *Security relates to the prevention of intrusion or confidentiality.*

The basic principle is to prevent the object of the security (the data) from entering a communication channel: to do so would be compromising for a certain party, and the system needs to be secured in light of that fact. Generally, we speak of a "gray channel" for the secured communication channel. Applications used to communicate over the gray channel must have a mechanism for encoding and decoding the data exchanged. This mechanism must be able to serve the security objectives.

The CENELEC EN 50159 standard [CEN 11b] defines different security objectives depending on the type of network. If we are dealing with a closed network, the protective measures relate to the integrity of the data received. With an open network, we need to take account of the possibility of unauthorized message transmissions[4] and modification of the devices.

In the final analysis, the basic errors relating to the messages are limited to:

– repetition: an existing message is re-sent at an inappropriate moment;

– deletion: a message is destroyed (e.g. an emergency brake command);

– insertion: a message is inserted into the existing data stream;

– sequencing: the sequence of messages is altered (e.g. a message might be delayed);

– corruption: a message is modified into a different correct message;

– delay: the transmission system is overloaded;

– masquerading: an intruder manages to pass for one of the interlocutors.

CENELEC EN 50159 proposes an approach to handle and demonstrate security in communications. Ultimately, one of the features of railway applications is that security is delivered by applications (rather than by the network) and that in case of failure, the system will enter a predefined security state (usually passively).

This explains why security (see Definition 3.9) is taken into account in the design of railway systems. Yet, as explained in [BOU 14b], the addition of

4 Unauthorized messages may be transmitted after eavesdropping on or monitoring communication channels. The eavesdropping or monitoring is not, in this case, the undesirable event.

remote access requires the taking into account of other types of risks, such as intrusion, hacking, control appropriation, etc.

3.3.2. *Safety management*

The safe operation [LIS 95, VIL 88] of a system is ensured by a certain number of activities based on studies. Such studies must be performed prior to the design of a system which needs to be safe.

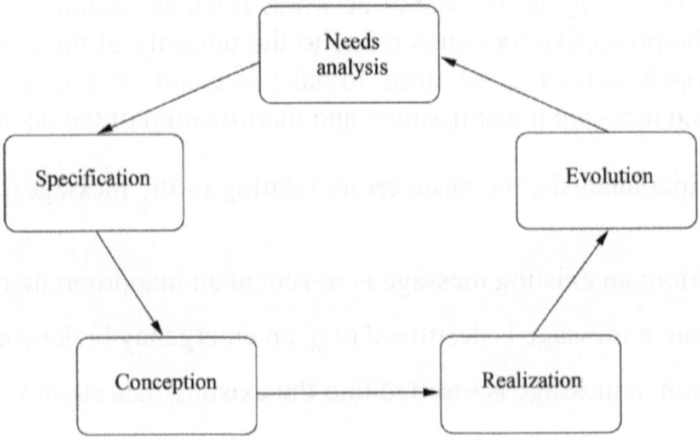

Figure 3.5. *Operational safety implementation cycle*

Figure 3.5 shows the cycle for implementation of operations safety for a computer system. The first step is to conduct a rigorous needs analysis. The studies relating to operational safety must be capable of identifying the risks and projecting their potential consequences. Thus, it is possible to define operational safety objectives (in terms of reliability, maintainability, availability and/or safety) for our system.

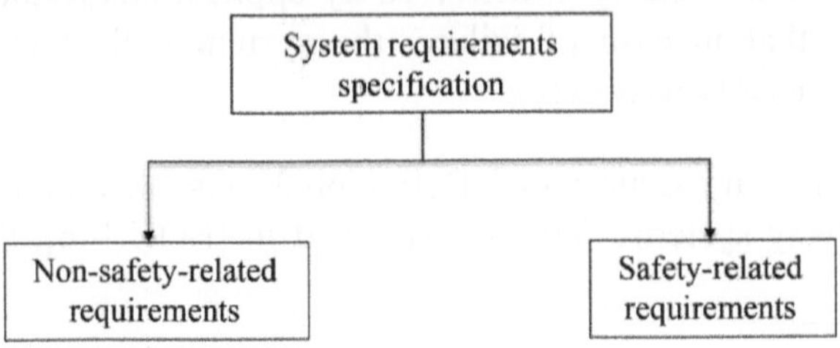

Figure 3.6. *Specification of system requirements*

The studies relating to operational safety follow the same phases as the lifecycle (specification, design, production and modification).

Specification is an important step, because it enables the designers to identify the needs and to clearly set out the system requirements. There are two types of such requirements: those not related to safety (functional requirements, non-functional requirements such as performance, etc.) and those related to safety (see Figure 3.6).

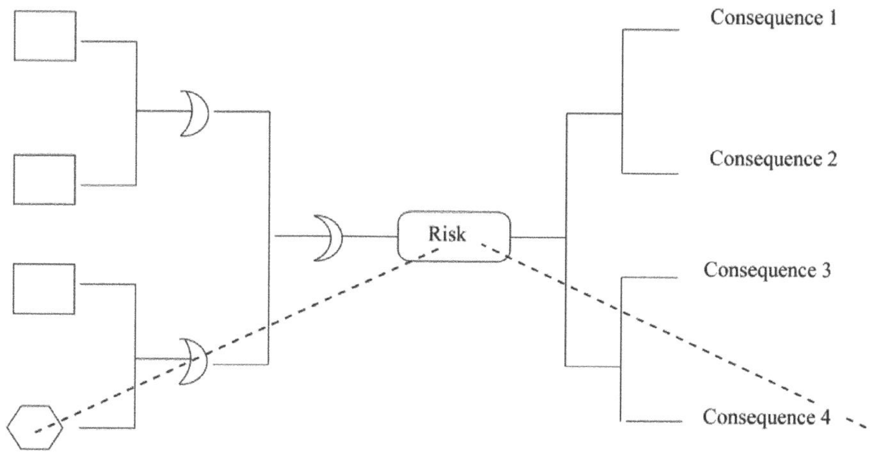

Figure 3.7. *Relationship between a failure, a risk and the potential effects*

As is shown by Figure 3.7, it is possible to form a link between a risk, the consequences associated therewith and the potential faults.

DEFINITION 3.10 (Risk).– *Risk is a combination of:*

– *the likelihood of occurrence of an event or series of events leading to a hazardous situation, or the frequency of such events; and*

– *the consequences of that hazardous situation.*

Thus, in light of Definition 3.10 and as specified by [IEC 08], risk is a combination of the probability of harm occurring and the severity of that harm. Severity is characterized by the nature of the harm and the consequences of the accident (see Figure 3.8).

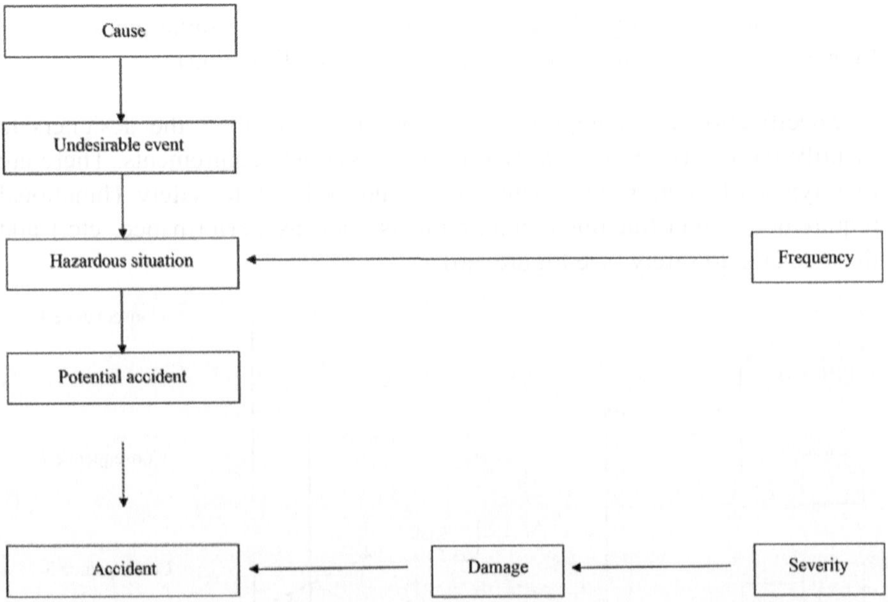

Figure 3.8. *Frequency and severity*

For each hazardous situation, therefore, we need to define a probability of its occurrence, or its frequency. With that goal in mind, it is possible to define a set of categories (see Table 3.2, which is taken from CENELEC EN 50126 [CEN 99]), where each category is associated with a frequency range.

Category	Description
Frequent	Likely to occur frequently. The hazard will be continually experienced.
Probable	Will occur several times. The hazard can be expected to occur often.
Occasional	Likely to occur several times. The hazard can be expected to occur several times.
Remote	Likely to occur sometime in the system lifecycle. The hazard can reasonably be expected to occur.
Improbable	Unlikely to occur but possible. It can be assumed that the hazard may exceptionally occur.
Incredible	Extremely unlikely to occur. It can be assumed that the hazard may not occur.

Table 3.2. *Frequency of occurrence of hazardous situation*

In Table 3.2, the term "often" indicates that the hazardous situation has occurred in the past and will likely occur again.

DEFINITION 3.11 (Hazardous situation – IEC 61508).– *A hazardous situation is a situation wherein a person is exposed to a hazardous phenomenon (or hazardous phenomena). A hazardous phenomenon is a potential source of danger in the shot and/or long term.*

Definition 3.11 introduces a link between the risk and the potential accident. The hazardous situation is one of the links in the chain in an accident scenario. It corresponds to an unstable, but remediable, state of the system.

For each hazardous situation, we need to analyze the consequences it will have for the system, the people involved and the environment. In order to quantify these consequences, we can define a level of severity. Table 3.3 is reproduced from CENELEC EN 50126, and forms the link with the consequences.

Given the diversity of risks and the impact that they may have on the system, it is preferable to define categories and associate actions with those categories. Table 3.4 is an example of categorization. In the undesirable category, there is a link with the supervisory authority. In the railway domain, this would be the rail operator or the regulatory authority.

Level of severity	Consequences for people or the environment	Consequences for the service
Catastrophic	Fatalities and/or multiple severe injuries and/or major damage to the environment	
Critical	Single fatality and/or severe injury and/or significant damage to the environment	Loss of major system
Marginal	Minor injury and/or significant threat to the environment	Severe system(s) damage
Insignificant	Possible minor injury	Minor system damage

Table 3.3. *Hazard Severity Level*

Risk category	Actions to be applied against each category
Intolerable	Shall be eliminated
Undesirable	Shall only be accepted when risk reduction is impracticable and with the agreement of the Railway Authority or the Safety Regulatory Authority, as appropriate
Tolerable	Acceptable with adequate control and with the agreement of the Railway Authority
Negligible	Acceptable with/without the agreement of the Railway Authority

Table 3.4. *Qualitative risk categories*

In order to evaluate the risk on the basis of the definitions given above, the standards recommend the establishment of a Frequency/Severity matrix. By way of example, here we have reproduced the matrix (Table 3.5) given in the CENELEC EN 50126 standard.

Frequency of occurrence of a hazardous event	Level of risk (risk category)			
Frequent	Undesirable	Intolerable	Intolerable	Intolerable
Probable	Tolerable	Undesirable	Intolerable	Intolerable
Occasional	Tolerable	Undesirable	Undesirable	Intolerable
Remote	Negligible	Tolerable	Undesirable	Undesirable
Improbable	Negligible	Negligible	Tolerable	Tolerable
incredible	Negligible	Negligible	Negligible	Negligible
	Insignificant	Marginal	Critical	Catastrophic
	Severity level of hazard consequence			

Table 3.5. *Qualitative risk category*

REMARK.– The content of Tables 3.2, 3.3 and 3.4 is taken directly from the published standards. The formulation of the elements takes account of the general nature of the possible application of the tables. Any time the tables are used, the first step must be to define the terms and their associated categories in the particular use context at hand.

The acceptable (or tolerable) risk is a value of a risk level arrived at by an objective, deliberate decision. This threshold of acceptability is known as the THR (Tolerable Hazard Rate). The THR is expressed as the probability of occurrence of a failure, expressed in the form 10^{-x} per hour. When identifying hazardous situations, it is crucial to assign them a THR, which must be included in the specification.

For a particular system, it is the responsibility of the rail operator to state the maximum THRs for each potential hazard. The process of safety management for the railway domain is defined by CENELEC EN 50126 [CEN 99], and is illustrated by Figures 3.4 and 3.5.

Risk management where the risks have the capacity to surpass the THR involves:

– decreasing the probability of occurrence by way of preventive displays to reduce the vulnerability of the elements of the system which are most exposed to that hazardous situation;

– decreasing the severity of the consequences by putting defensive measures in place.

On the basis of an initial description of the system and a list of the hazardous situations, it is possible to perform a risk analysis, aimed at identifying which measures need to be taken in order to render the risks acceptable. These measures need to be taken when creating the system.

Figures 3.9 and 3.10 show an example of the process of risk analysis. This process is applicable to the system, the subsystems and the devices. The phase of identification of the hazardous situations consists of using the system interface to study the identification of hazardous situations but also defining the THR associated therewith.

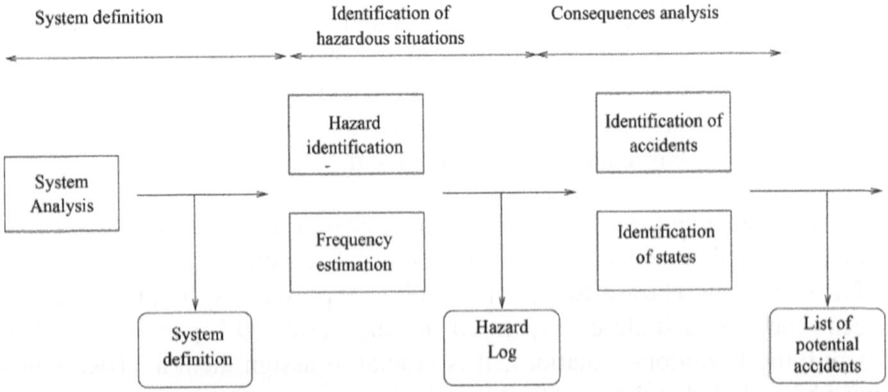

Figure 3.9. *Example of process of risk analysis – Part 1*

With regard to the risks, if it is possible to eliminate the danger, obviously we should do so, because this is the most effective way of reducing the risks. Yet it is often impossible to eliminate the dangers. In such cases, we need to reduce the risks to an acceptable level.

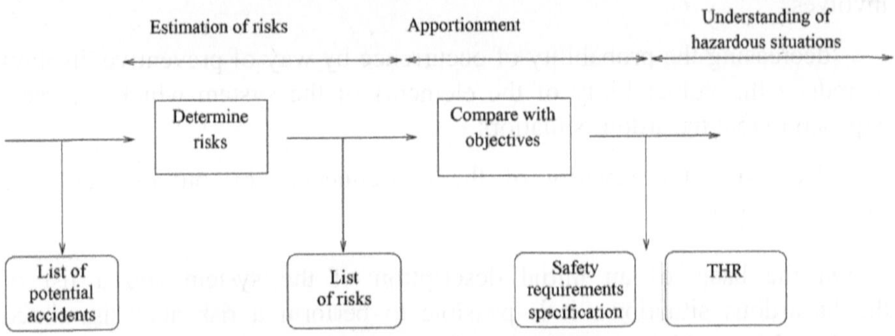

Figure 3.10. *Example of process of risk analysis – Part 2*

In order to reduce the risks, we need to reduce the frequency of the danger and/or reduce the severity of the danger (i.e. the consequences). The reduction of the frequency and/or severity of the danger is more commonly denoted by the term "risk management".

Figure 3.11 presents a diagram of the process of risk management, which is an iterative 3-phase loop: identifying, evaluating and acting.

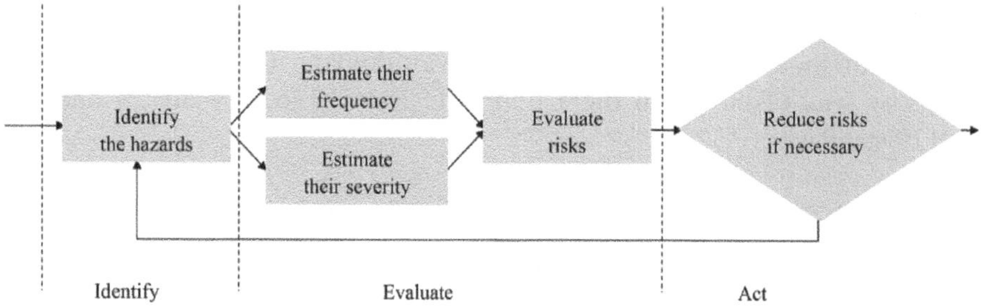

Figure 3.11. *Risk-management process*

The objectives of risk management are:

– step 1: to identify the dangers associated with the system:

- identify any systematically hierarchically rank all the reasonably-predictable dangers associated with the system in its application environment,

- identify the sequences of events which lead to the dangers;

– step 2: evaluate the dangers:

- estimate the frequency of occurrence of each danger,

- evaluate the probable severity of the consequences of each danger,

- evaluate the risk for the system posed by each danger;

– step 3: establish a process to manage the risk at hand:

- establish the hazardous situations register[5] (HSR).

Based on the list of safety requirements and the THRs, it is possible to define a safety integrity level.

3.3.3. *Safety integrity*

In the various domains (aeronautics, nuclear, space, railways, etc.), a similar concept is found: that of the safety level. Yet the concept of safety integrity was first introduced by the IEC 61508 standard [IEC 08].

5 We also speak of a safety log and/or HazardLog (HL).

The concept of safety integrity was then taken up and adapted in the various offshoot standards. For the railway domain, the concept of safety integrity was taken up in the CENELEC EN 50126, EN 50128 and EN 50129 standards.

The concept of safety integrity comprises two components:

– integrity against random failures.

– integrity against systematic failures.

The main difference between random and systematic failures lies in the fact that systematic failures cannot be quantified by way of probabilistic computation.

Random failures result from the hardware; they occur randomly over time, and are due to aging and wear-and-tear on the hardware. Because of their nature, software applications are not subject to random failures.

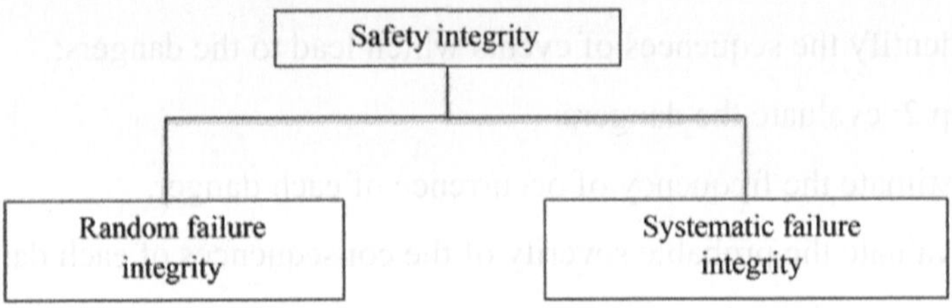

Figure 3.12. *Safety integrity*

Systematic failures may be due to the hardware and/or the software. They are due to human errors during the different phases in the life of the system (a specification and/or design problem, a manufacturing problem, a problem during the installation, a problem caused during maintenance, etc.). With regard to systematic failures, it is necessary to employ various methods or techniques which should help obtain a sufficient level of confidence in the safety integrity level attained.

Figure 3.6 presents the division of the requirements specification into two categories: safety-related requirements and non-safety-related requirements. In view of what was said above, it is possible to refine this division, as illustrated by Figure 3.13. This figure introduces a new division of the safety-related requirements into functional safety requirements and safety integrity requirements.

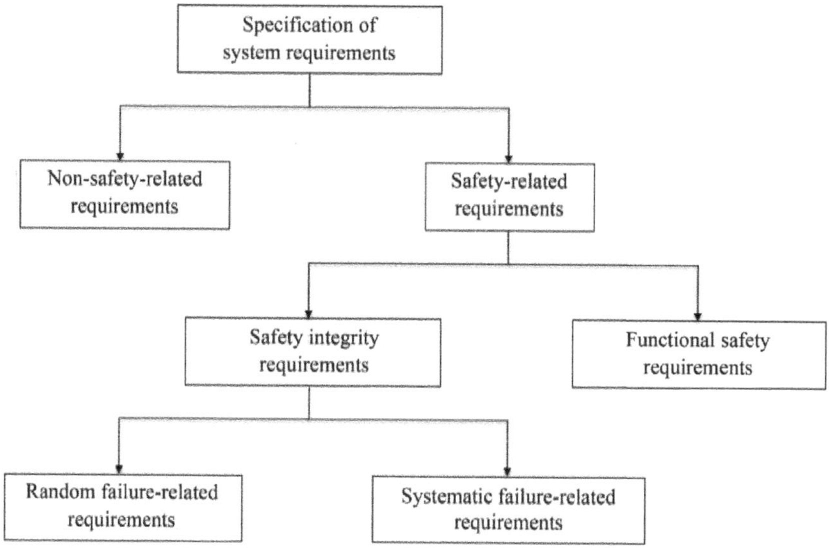

Figure 3.13. *Specification of system requirements*

The functional safety requirements characterize the safety-related requirements linked to the functionality of the system's safety functions.

The concept of safety integrity can be characterized as the ability of a system to fulfill its safety-related functions in specific conditions, in a predefined operational environment for a given length of time. Safety integrity (section 4.7 in CENELEC EN 50126 [CEN 99]) is linked to the probability of not performing the required safety function.

Unlike with IEC 61508 [IEC 08], CENELEC EN 50126 (which is applicable to the whole of the railway system) does not define the correlation between safety integrity and the probabilities of failure. On the other hand, it is necessary to define that correlation for a given application. The safety functions are assigned to safety-related systems or to external risk-reduction devices.

The SIL is divided into four discrete values (1 to 4) which can be used to specify the prescriptions relating to safety integrity.

For IEC 61508 [IEC 08], the four SILs for a safety-related function are characterized by the impact of any failures:

– SIL 4: catastrophic impact (highest level);

– SIL 3: impact on the community;

– SIL 2: major protection of the installation and of production, or risk of injury to employees;

– SIL 1: minor protection of the installation and of production (lowest level).

The SIL must be determined by experts, and the SIL must be assigned to an element which, alone, is capable of serving one or more functions.

3.3.4. *Determination of the SIL*

It is necessary to put a systematic approach in place (see Figures 3.9 and 3.10) to the determination of the safety requirements. This approach must take account of the environment, the modes of use and the architecture of the element under analysis.

After having identified the risks and hazardous situations needing to be taken into account, we have a list of the *hazards* and the THRs associated therewith.

Appendix A in CENELEC 50129 [CEN 03] introduces the method of subdivision of the THR assigned to each undesirable event. The proposed method can be broken down into five steps:

– step 1: analysis of the causes of each undesirable event in order to identify the system functions whose failure leads to the undesirable event;

– step 2: division of the THR allocated to the undesirable event into THRs allocated to the system functions using AND/OR combinatorial logic;

– step 3: allocation of an SIL to the system functions using the corresponding SIL table (see Tables 3.6 and 3.7);

– step 4: division of the THR and the SIL of each system function into the different subsystems supporting that function.

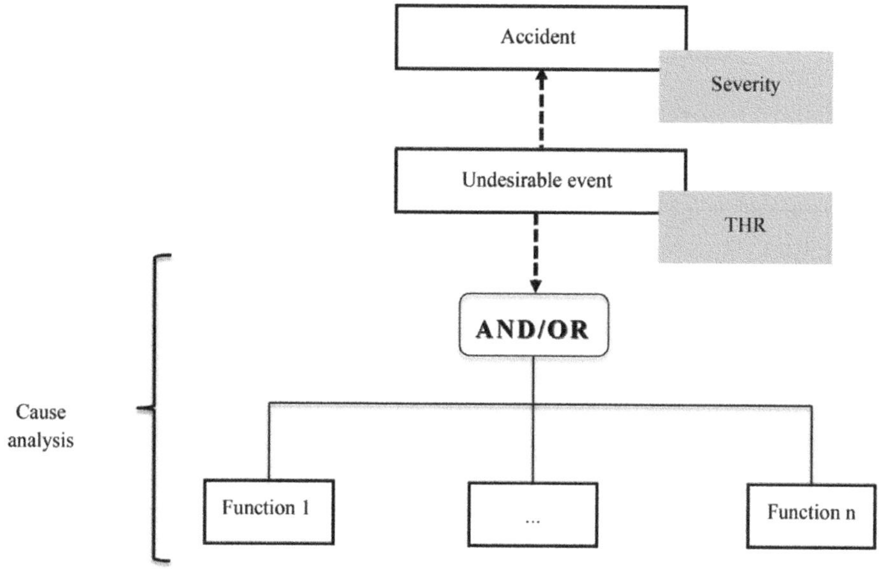

Figure 3.14. *Analysis of causes associated with each undesirable event*

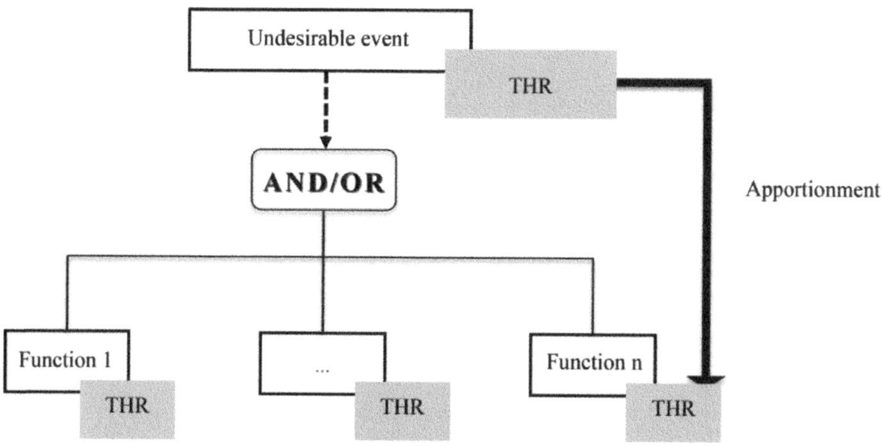

Figure 3.15. *Allocation of THRs as a function of the fault tree diagrams*

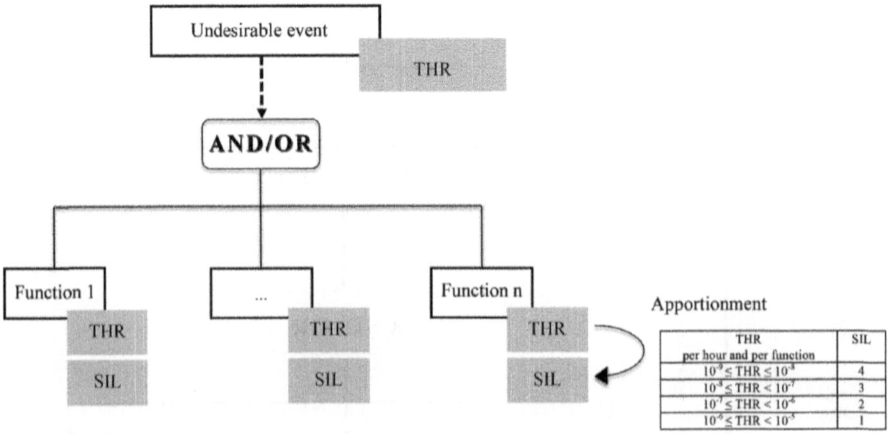

Figure 3.16. *Allocation of SILs of the basis of the THRs*

The left side of Figure 3.17 shows the architecture analysis whose input is a given platform, the list of system functions and their THRs, which analyzes the link between the system functions and the subsystem functions.

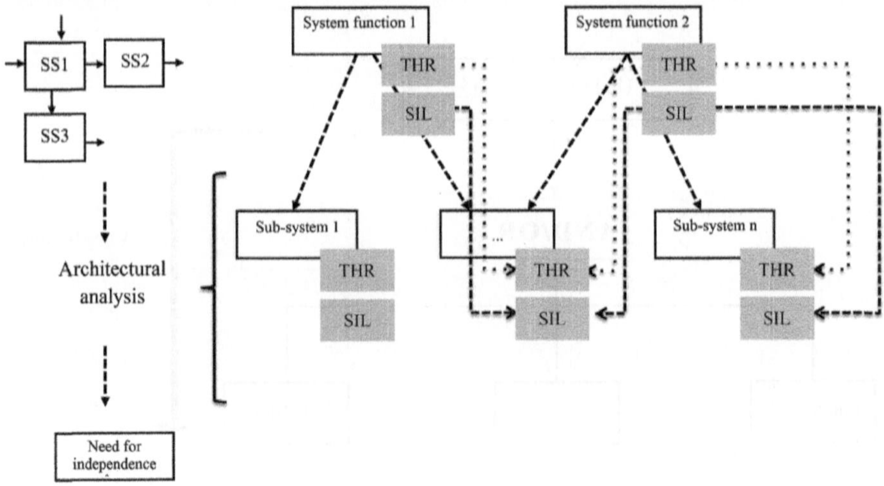

Figure 3.17. *Allocation to the subsystems*

On the right-hand side of Figure 3.17, we show the link between the system functions and the subsystem functions. If a subsystem function is used only by a system function, then there is a direct allocation of the THR and the SIL, and similarly in regard to the systemic and subsystemic levels. If a subsystem function is the support for several system functions, the SIL of the system function is equal to the maximum of the SIL of the subsystem functions and the THR of the system function is equal to the minimum of the THR of the subsystem functions.

An analysis of the causes of the failures will be used as the basis for this allocation; the THRs of the subsystem functions will be calculated using the fault tree diagram. As regards the allocation of the SILs:

– the functions, the impact on safety of whose failures are combined by way of an "OR" branch in a cause tree, are allocated the same SIL as that which was determined prior to the "OR" branch;

– the allocation of SILs to the functions, the impact on safety of whose failures are combined by way of an "AND" branch in a cause tree, depends on the respect of criteria of functional and physical independence of those functions. If the functional and physical independence of the failure modes of the functions is clearly established, the SIL which was determined prior to the "AND" branch can be distributed amongst the functions situated on both sides of that "AND" branch. Otherwise, the prior SIL must be allocated to the functions;

– step 5: subdivision of the objectives: the THR of each subsystem is divided into Failure Rate (FR) of the hardware platforms comprising the subsystem depending on the architecture (independence hypotheses may then be made). The overall SIL is divided into SILs in each subsystem in SILs (SSIL) for the hardware platforms (or software):

- the SIL of the hardware is identical to the SIL of the subsystem;
- the Software SIL (SSIL) is identical to the SIL of the subsystem.

It should be noted that we adopt the hypothesis that a subsystem can be divided into a "Hardware" and "Software" part.

The switch from the subsystem SIL to local equipment SIL is made with rules that were introduced, but it is possible to take account of the architecture of the equipment with a view to refining the SILs and SSILs.

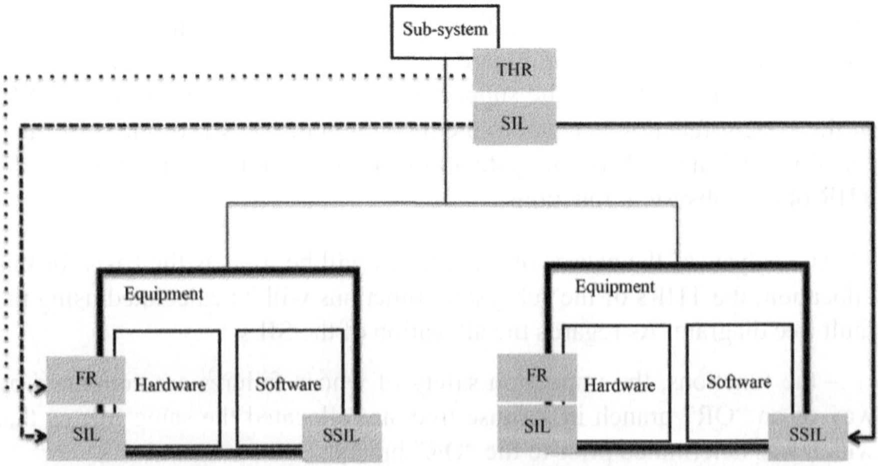

Figure 3.18. *Division of the SILs and THR subsystem between the hardware platforms*

Indeed, if we implement a homogeneous 2oo3 architecture (same hardware elements, same OSs[6], same memories, etc.), we can improve the taking into account of random faults, but not design faults or common faults (processor faults, etc.), so therefore the SIL required for the hardware needs to be the same as that of the subsystem.

If we put the same software on the same architecture three times, the software then needs to be SSIL4. If we are able to diversify and demonstrate the diversification of the software applications, the implementation of three diversified applications can justify an SSIL2 for those applications.

Putting heterogeneous 2oo3 architecture (three different processors, diversified memory units, etc.) in place enables us to improve the handling of design problems and faults with the basic components, whilst limiting common modes. It is then possible to show that each processing unit can be assigned an SIL2 objective. If we put in place only one software application, it will remain SSIL4.

The reasoning processes presented above by way of example show that it is possible, on the basis of the architecture and an expert judgment, to refine

6 OS for Operating System.

the allocation of the SILs and SSILs. It should be noted that this allocation remains subject to an expert judgment, which must be justified, and that it is by no means an SIL algorithmic technique.

3.3.5. *SIL table*

IEC 61508 [IEC 08] defines two categories of systems: low-demand systems and high-demand systems. In order to characterize these two types of system, it identifies the concept of average probability of failure on demand (PFDavg) and the concept of probability of failure per hour (PFH). Table 3.6 presents the definitions of SIL for the two types of system.

Low demand	
Safety Integrity Level	Average probability of failure to execute the function on demand
SIL 4	10-5 ≤ PFDavg < 10-4
SIL 3	10-4 ≤ PFDavg < 10-3
SIL 2	10-3 ≤ PFDavg < 10-2
SIL 1	10-2 ≤ PFDavg < 10-1

High demand	
Safety Integrity Level	Probability of dangerous failure per hour
SIL 4	10-9/h ≤ PFH < 10-8/h
SIL 3	10-8/h ≤ PFH < 10-7/h
SIL 2	10-7/h ≤ PFH < 10-6/h
SIL 1	10-6/h ≤ PFH < 10-5/h

Table 3.6. *SIL table in IEC 61508 [IEC 08]*

For the railway domain, Table 3.7 shows the link which exists between the SIL and the THR. In fact, this table was introduced for railway signaling and can be extended to all or part of the system. This table can also be redefined.

THR per hour and per function	SIL
$10^{-9} \leq THR \leq 10^{-8}$	4
$10^{-8} \leq THR < 10^{-7}$	3
$10^{-7} \leq THR < 10^{-6}$	2
$10^{-6} \leq THR < 10^{-5}$	1

Table 3.7. *SIL table from CENELEC EN 50129 [CEN 03]*

REMARK.– It should be noted that Tables 3.6 and 3.7 should be read from left to right. Thus, a THR enables us to define an SIL. The SIL corresponds to a level of confidence which must be achieved as to the correctness of the system. With rare exceptions, it is possible to convert from an SIL to a THR. In order to do so, we express the SIL of a hardware element which is introduced to serve the SIL objective. As this hardware element must be quantified, a THR is then necessary.

3.3.6. *Allocation of SILs*

We have presented the general method for allocating SILs and SSILs in the previous sections. However, as indicated in the explanations given in step 5 in section 3.3.4, it is important to exercise caution: the choice of an SIL and an SSIL is one which is made by an expert on the basis of the risks identified, the safety studies (preliminary risk analysis, FMEA, fault tree analysis (FTA), SEEA, etc. – for a description of the studies performed, see [VIL 88]), objectives such as the THR and the proposed architecture.

There are approaches (Table 3.8) which are intended to define general rules for architectures, such as in the Yellow book[7] [RSS 07].

Other approaches, such as [HAR 06, PAP 10], go a little further, because they do not rule out the possibility of an automated SIL-allocation process.

SIL of global level	SIL of low-level functions		Combination function[8] (e.g. with voters)
	Primary	Other	
SIL4	SIL 4 SIL 4 SIL 3	None SIL 2 SIL 3	None SIL 4 SIL 4
SIL 3	SIL 3 SIL 3 SIL 2	None SIL 1 SIL 2	None SIL 3 SIL 3
SIL 2	SIL2 SIL 1	None SIL 1	None SIL 2
SIL 1	SIL 1	None	None

Table 3.8. *SIL attribution table, extracted from the Yellow Book [RSS 07]*

7 The Yellow Book describes the implementation of CENELEC EN 50126, EN 50129 and EN 50128 for the United Kingdom.

8 As a general rule, the combination function (e.g. a vote situation) has the same SIL as the global function, but its complexity is very low.

3.3.7. *SIL management*

Just as with the IEC 61508 standard [IEC 08], the CENELEC EN 50126, EN 50128 and EN 50129 standards detail the requirements which must be satisfied in order to achieve each safety integrity level. These requirements are stricter for higher SILs, so as to ensure a lower probability of a dangerous fault.

As Figure 3.19 shows, safety integrity is governed by two groups of practices. With regard to integrity against random failures, it is necessary to balance quantitative objectives (such as the *Failure Rate*) and qualitative objectives. As regards integrity against systematic failures, we must balance QA objectives, safety management, and objectives pertaining to the safety management of the hardware elements.

We shall now give a brief discussion of the techniques used to manage a platform's integrity against systematic failures:

– with regard to the measures employed to manage the safety of the hardware elements (see IEC 61508 Part 2; CENELEC EN 50129), they are based on safety analysis methods (PHA, FMEA, Hazard and Operability study (HAZOP), FTA, etc.), design methods[9] (redundancy, diversification, etc.), downgraded-mode and definition-strategy management and a command of the technical documentation;

– with regard to the quality management strategies (see IEC 61508 Part 1, CENELEC EN 50129 and CENELEC EN 50128), they are based on an understanding of quality (implementation of ISO 9001:2008 [ISO 08] or equivalent), understanding of staff certification and implementation of independence;

– with regard to the quality management strategies (see IEC 61508 Parts 1 and 3, CENELEC EN 50126, EN 50129 and EN 50128), they are based on an overall understanding of safety (definition of a strategy which covers everything from system level to the hardware platforms and the software they run), a demonstration of safety, management of the safety of the software applications (use of SEEA[10], time-honored development

9 For further information, see Chapter 1 of [BOU 10a].

10 Software Error Effects Analysis (SEEA) is the only safety study performed on all or part of a software application (see [AFN 90]).

methods and "safe" programming techniques) and an understanding of the technical documentation associated with the different activities (the activities must be auditable).

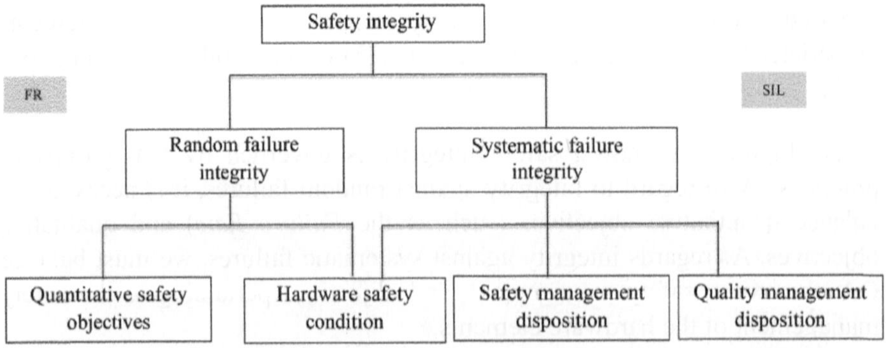

Figure 3.19. *Safety integrity management*

3.3.8. *Software SIL*

As indicated in section 2.2.2, there are four Safety Integrity Levels (SILs). When an element is not subject to the SIL standard, it is not associated with any level, which means that the standard is not applicable to that element.

For the software, on the other hand, the Software Safety Integrity Level (SSIL) has been defined. This scale has five values: 0 to 4. Level 4 corresponds to the highest safety integrity level, and level 1 to the lowest. Level 0 indicates that the particular piece of software is unrelated to safety.

The main difference between the SIL and the SSIL is that for SSIL 0, the CENELEC 50128 standard [CEN 01a, CEN 11a] identifies obligations. It should be noted that in the 2011 version of CENELEC 50128, the term "SSIL" is replaced by "SIL", but it is preferable to continue to use "SSIL" when speaking of software.

Note that CENELEC 50128 is written in such a way that one can consider there to be only three levels:

– SSIL 0: the software is not safety-related, but it is necessary to demonstrate quality assurance (QA) and configuration management;

– SSIL 2: the software is associated with a "medium"-level safety objective, which requires the implementation of production rules designed to guarantee safety;

– SSIL 3-4: the software is associated with a "high"-level safety objective, which necessitates the use of appropriate resources and techniques.

Normally, SSIL1 is not used because it is related to reliability (impact on hardware and low impact on humans). It is not a good idea to try to manage software reliability because it is not possible to estimate the number of residual software defects, and each time we update the software we add new defects and/or cause residual effects.

CENELEC EN 50128 [CEN 01a, CEN 11a] identifies the tools and methods designed to produce a "0-fault" program. Below is an extract from the list of techniques recommended by CENELEC EN 50128:

– use of formal methods;

– use of dynamic tests;

– use of qualifiable development environments;

– use of simulation methods to validate the model and/or select the tests to be run;

– formalization of teamwork methods (specification, encoding, verification, validation, etc.);

– implementation of strong quality assurance (using a pre-established, disciplined system).

3.3.9. *Iterative process*

The process of definition of the SILs which starts with the hazards and runs down to the lowest levels (software are hardware) is a process which must loop back on itself, as is illustrated by Figure 3.20. The purpose of this loop is to check that all the predefined objectives have been achieved.

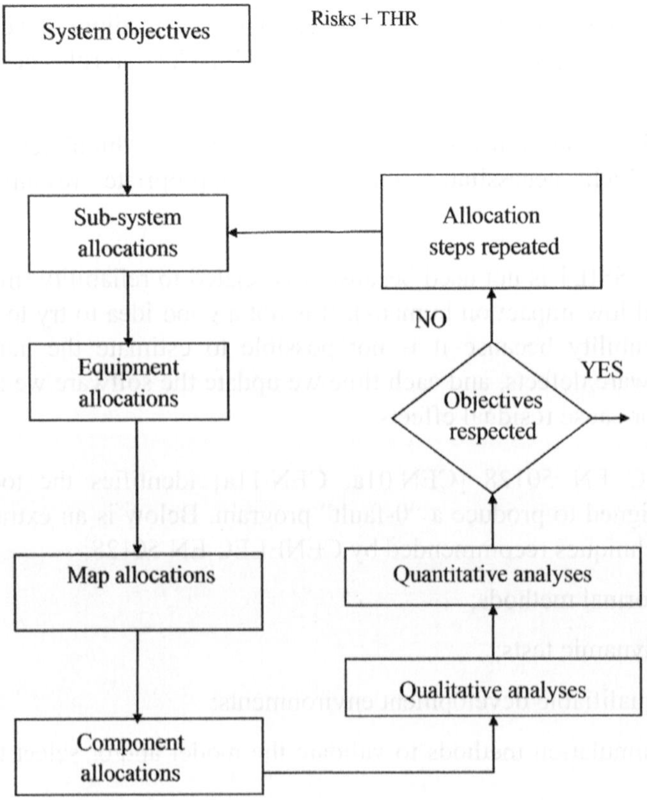

Figure 3.20. *Allocation of the system to the fundamental elements*

3.3.10. *Identification of safety requirements*

Figure 3.13 introduced the safety-related requirements, but did not indicate how the link between the risk analyses and the safety requirements is established. A grasp of the risks enables us to identify the barriers enabling us to work through to the potential accident and/or to decrease the severity of the consequences of the accident.

Figure 3.21 presents the barriers in the series of events leading up to the accident. As a general rule, the barriers relate to the equipment and the system functions and/or the implementation of special operational procedures.

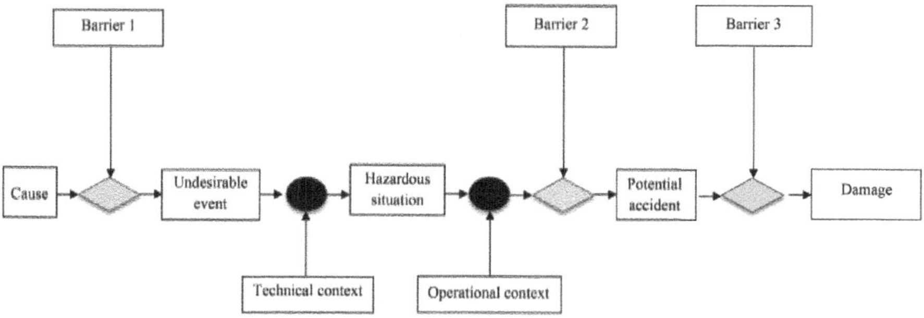

Figure 3.21. *Introduction of barriers*

Safety studies should enable us to define the safety requirements which will be served by the barriers. Figure 3.22 shows a certain number of elements (gray boxes) which could offer the support for safety requirements.

The process began with the identification of the risks and we were able to identify the system functions relating to safety, and to divide those system functions between the devices, in order to identify the devices that support the safety-related functions. Finally, we associate with each device a set of requirements which is a function of the risk incurred.

One of the important points for this type of analysis is the identification of the choices and justifications.

The SIL can then be defined as the level of confidence accorded to the respect of the safety requirements allocated to the function and/or the device.

3.4. In IEC 61508 and IEC 61511

[ISA 05] offers an interpretation and certain rules for the implementation of IEC 61508 [IEC 08] and IEC 61511 [IEC 05].

IEC 61508 [IEC 08], in Part 5, describes the risk graph as a means of identifying the SIL. IEC 61511 [IEC 05], in Part 3, presents several methods for calculating the SIL such as, for instance: the risk graph and the Layer of Protection Analysis (LOPA). For further information about these methods, readers could usefully consult [GUL 04] and [AHM 09].

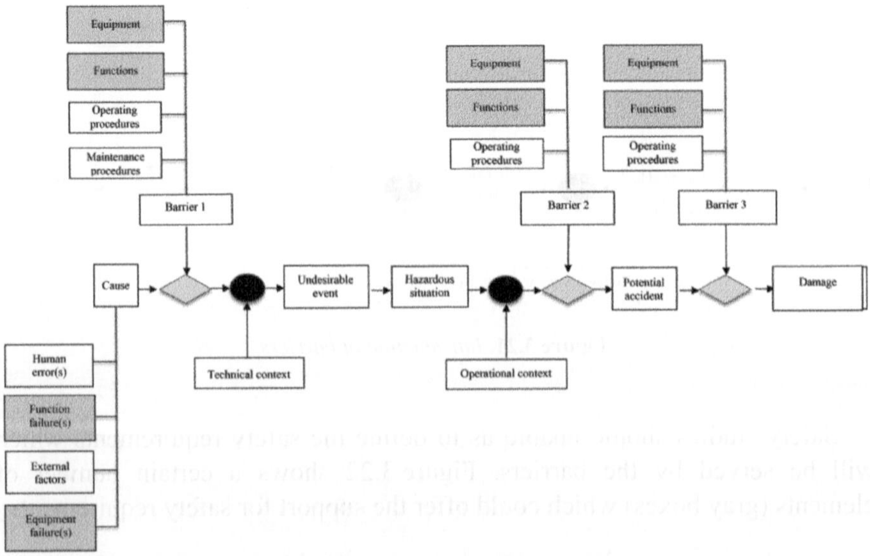

Figure 3.22. *Position of safety requirements*

These three methods to determine the SIL for a given safety instrumented function need to be performed after a risk analysis is conducted. It should be noted that, unlike with Appendix A of CENELEC EN 50129 [CEN 03], which is normative, Appendices D, E and F of Part 3 of IEC 61511 (and for IEC 61508) are informative – thus, it is not obligatory to use these methods.

3.4.1. *Risk graph*

One of most widely used techniques to determine the SIL in safety instrumented systems (SIS) is the so-called "risk graph" method (see IEC 61508 [IEC 08] and IEC 61511 [IEC 05]).

The adoption of this method leads to the introduction of a certain number of simplifying parameters to describe the nature of the hazardous situation when the safety-related systems are faulty or unavailable.

As indicated above, the risk "R" is seen as the combination of a frequency "f" and a level of severity – we tend to speak more of the consequence, and therefore use the symbol "C". The frequency of the hazardous event (f) is supposed to be the result of three influencing factors:

– frequency and length of exposure in a hazardous area;

– the possibility of preventing the hazardous event;

– the probability that the hazardous event will occur in the absence of safety-related systems, which is sometimes called the probability of an undesired occurrence.

If we combine the fact that R = f * C and that f = F * P * W, then we can characterize the degree of risk by four parameters:

– the consequence of the hazardous event (C);

– the frequency and length of exposure to danger (F);

– the possibility of preventing the hazardous event (P);

– the probability of the undesired occurrence (W).

For a particular risk, therefore, we simply need to analyze each parameter and associate them in order decide on the SIL for the safety-related systems.

These four parameters are considered to be sufficiently generic to be applicable to all systems, and they can be used to create a meaningful hierarchical ranking of the risks. It is possible to introduce new parameters in order to refine the methodology.

Figure 3.23 shows an example of a risk graph. In the scales associated with the parameter "W", we see the emergence of six classes of requirements, graded from "a" to "b", through SIL1 to SIL4. The category "a" thus corresponds to "no safety requirement" and category "b" to "situation unacceptable".

For a specific branch of the risk graph (i.e. X1, X2, ... or X6) and, for a specific W scale (i.e. W1, W2 or W3), the final output of the risk graph gives the SIL for the system under analysis (1, 2, 3 or 4) and corresponds to a measurement of the necessary reduction in risk for the system.

In view of the possibility to adapt the parameters to the specific domains and professions involved in the company, the main difficulty lies in the calibration of the graph. The ranking of the parameters thus needs to take account of all situations.

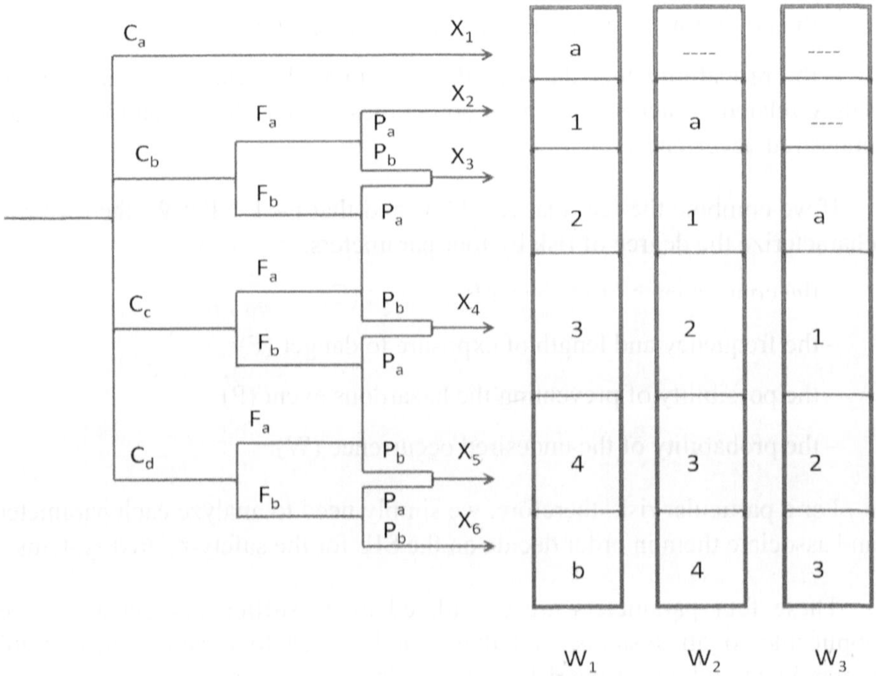

Figure 3.23. *Example of risk graph*

One of the difficulties then is to take account of the imprecision and of uncertainty on the part of the experts; for this purpose, fuzzy-logic-based approaches have been put forward (for instance, see [SIM 06]).

3.4.2. *LOPA*

As indicated in IEC 61511 [IEC 05], the LOPA method (for Layer of Protection Analysis – see [AIC 93]) begins with the data gathered during the HAZOP analysis, and takes account of each hazard identified, documenting the triggering cause and the layers of protection which prevent or lessen the hazard. Unlike what happens with purely qualitative risk-assessment techniques, LOPA gives an estimation of the frequency of an undesirable event.

The LOPA method evaluates the reduction of the risk by analyzing the contribution of the different layers of protection (from the intrinsic characteristics of the process to any emergency measures that are in place) in case of an accident. It is used to determine which SIL should be allocated to each safety instrumented function (SIF) and how many layers of protection are necessary to reduce the risk to an acceptable level. The aim is to calculate the residual risk expressed in terms of frequency of accidents per year, which necessitates that we quantify the frequencies of occurrence of triggering events and the probabilities of failure in each layer.

The analysis is made up of the following steps:

– definition of the impact of the undesirable event (severity);

– determination and enumeration of all triggering events;

– determination and enumeration of all the layers of protection preventing the propagation of the triggering event leading to the undesirable event;

– determination of the frequency of the triggering events, based on data or on expert assessments;

– determination of the effectiveness of the layers of protection in terms of probability of failure when needed;

– calculation of the frequency of the undesirable event.

The LOPA method applies only for function on demand (the safety system is called upon only in the presence of an event triggering the hazardous situation which is independent of that event) and it is not suitable for continuous mode (any fault in the safety system is a triggering event for the hazardous situation).

Unlike with the risk graph method, the LOPA method has the advantage of not requiring scaling, because the input values are quantified. The problem does arise, however, with those values which are not commonly accepted and which differ depending on the sites, the situations, the feedback from the ground, etc.

3.4.3. *Overview*

One of the main difficulties introduced by this type of approach is the need to validate a certain number of parameters which have a direct impact on the choice of the SIL.

The use, in a railway application, of products whose SILs are assigned with this type of approach will thus be faced with the difficulty of demonstrating that the initial hypotheses are verified by the use context.

3.5. Conclusion

In this chapter, we have presented the basic method for SIL allocation as it applies in the field of railway signaling.

As a general rule, we apply the same approach for the other railway subsystems, but the IEC 61508 framework [IEC 08] is being used increasingly frequently for the auxiliary parts (tunnel ventilation management system, PA system on the platforms, etc.).

By defining a normative framework describing what a railway system is, the aim in the railway domain is to define a railway system as a collection of products. Building a railway system could, in this way, be viewed as being similar to building using LEGO®, but with the need to verify the compatibility of the products.

The acceptance of previously-certified products is therefore a point of crucial importance, which requires the use of cross-acceptance analyses.

4

Software Assurance

4.1. Introduction

In this chapter, we shall present the prerequisites necessary to implement CENELEC 50128. The 2011 version of CENELEC 50128 [CEN 11a] introduces the concept of software assurance (SwA). SwA covers all activities leading to the confidence that the software is of a certain software safety integrity level (the SSIL mentioned previously).

For the *hardware* aspects, there is a whole collection of techniques (redundancy, diversity, etc.) which can be employed to create safety equipment that takes account of faults in the basic elements (memory, processor, ALU, etc.). However, for software, such an approach is unusable. After all, if we insert safety measures into the software program, we inevitably make it much more complex, so the number of systematic latent faults increases, because it is difficult to adequately check very complex software.

4.2. Prerequisites

CENELEC 50128:2011 considers the implementation of ISO 9001:2008 [ISO 08b] as a prerequisite; so too does CENELEC 50126 [CEN 99], pertaining to safety management. For this reason, CENELEC 50128:2011 does not identify any activity or organizational constraint for quality management or safety management.

4.3. Quality assurance

4.3.1. *Introduction*

Definition 4.1 introduces quality assurance with an emphasis placed on an objective: confidence in the attainment of a prerequisite level of quality.

DEFINITION 4.1 (Quality assurance – QA*).– Implementation of an appropriate set of pre-established, systematic measures to ensure confidence in the attainment of the required level of quality.*

Although its role is to be helpful, quality is often seen by the project participants as:

– a "word": it is useless, it is not a tangible part of the project, it contributes nothing, etc.;

– a "curse": it is a waste of time, it creates more problems than it solves, it is a straitjacket, etc.

Quality assurance is an exercise in prescription and control which must be understood and accepted. It must provide methods and processes, and must facilitate control of the activities. Thus, the implementation of quality assurance enables us to have pre-established, systematic activities which are clearly understood and mastered.

The general measures taken by a company to ensure the quality of its products or services are described in that company's "Quality Assurance Manual" (QAM). Each measure (specification, test, etc.) is defined as part of a "procedure". In a complex prescribed strategy, a set of guidelines may be available, describing how to implement that strategy. Each procedure associated with a strategy identifies the input and output documents. Typical guidelines describing the template of the required documents should also be available.

By employing the pre-established, systematic strategies prescribed in the quality framework (QAM, procedure, template), skill and efficiency begin to be seen. Indeed, skill is gained by application and understanding of the processes. Efficiency, for its part, is achieved by the implementation of proposed improvements, the understanding and acceptance of the difficulties encountered during the process, and the implementation of feedback procedures.

4.3.2. *Quality assurance management*

4.3.2.1. *Prescriptions of standard in force*

CENELEC 50128:2011 reiterates the need to have a quality-assurance strategy based on ISO 9001:2008. In previous versions of the standard, the existence of a QA system at company level is mandatory (M), and ISO 9001:2008 certification is simply highly recommended (HR), but as Table A.9 in the latest version of the standard (reproduced here as Table 4.1) shows, conformance with ISO 9001:2008 has become mandatory. Table A.9 is applicable to all entities involved in the creation of the software, and hence also extends to the subcontractors and/or co-contractors.

Conformance of the quality strategy to ISO 9001:2008 needs to be demonstrated. With this in mind, it is possible to establish a matrix (see Table 4.2, for example) which, on each count of the ISO 9001:2008 requirements, illustrates the link with the company's quality system (QAM, procedure, professional guidelines, etc.).

Measurement	SSIL0	SSIL1; SSIL2	SSIL3; SSIL4
Accredited EN ISO 9001	R	HR	HR
Compliant with EN ISO 9001	M	M	M
Compliant with ISO/IEC 90003	R	R	R
Company's quality system	M	M	M
Software configuration management	M	M	M
Checklist	R	HR	HR
Traceability	R	HR	M
Data recording and analysis	R	HR	M

Table 4.1. *Reproduction of Table A.9 from CENELEC 50128:2011*

Table 4.1 shows new requirements: it is HR to use control lists and a traceability strategy[1] (traceability is mandatory for SSIL3-SSIL4). These two requirements are not new; indeed, it is not reasonable to attempt to verify an activity without explicitly stating an objective for that activity. Control lists are one means to set a direction, as the control list can take account of

1 Requirement management and traceability are presented in Chapter 5.

various guidelines linked to different departments (quality, V&V, technical and safety).

ISO	Description Status
4.1 General Requirements	Partial
4.2.1 General Documentation Requirements	Partial
4.2.2 Quality Manual	Partial
4.2.3 Control of Documents	Total
4.2.4 Control of Records	Total
5.1 Management Commitment	Partial

Table 4.2. *Conformity matrix to ISO 9001:2008*

As for traceability, it is the hinge point of the whole of the CENELEC referential framework: it is necessary to identify the functional and non-functional requirements on the basis of the system requirement specifications (SyRS), extending to the hardware and/or software elements.

In the railway domain, the software will invariably be subject to independent assessment. With this in mind, it is necessary for the activities to be identically reproducible and that records are kept which are auditable[2] for an external body. Hence, it is important to record and analyze the elements produced at each step, which is why the last item in Table A.9 is present, rendering data-recording and analysis "highly recommended" even when the SSIL is 0. Independent software assessment is discussed in Chapter 10 of this book.

4.3.2.2. *ISO 9001:2008*

The international standard ISO 9001:2008 [ISO 08b] is a generic norm which specifies the requirements pertaining to the quality management system (QMS) in an organization. It is used when an organization needs to demonstrate its capacity to regularly deliver a product which conforms to the customer requirements and to the applicable legal and regulatory

2 "Auditable" means that an external body is able to perform an audit *a posteriori* and the elements produced are sufficient to judge whether or not the activity being audited has been correctly carried out. Thus, it is necessary to leave sufficiently precise records of the activity as it is being performed.

requirements. It also aims for the continuous improvement of the system and assurance of conformity with those requirements.

The ISO 90003 guide [ISO 04a], although it takes account of the version of ISO 9001 from 2000, still remains a useful application guide for the application to software development but it is not an obligation to implement CENELEC 50128:2011 and IEC 62279.

All the requirements outlined in ISO 9001:2008 can be applied to any and all organizations, regardless of their type, size and the final product in question.

The main advantage of this standard is that it necessitates the definition of the professional processes so as to make them universally applicable. Thus, they constitute the knowledge of the company and ensure replay[3] and maintainability over time.

ISO 9001:2008 comprises five main chapters:

– Quality Management System (process approach, general requirements, documentation) – continuous improvement of the QMS;

– responsibility of the Directors (responsibility, authority and communication, quality policy, quality objectives, directors' review);

– resource management (human resources, infrastructure, working environment);

– product development (planning, client process, purchasing, production, measuring equipment);

– measurement, analysis and improvement (control, non-conforming products, internal audits, corrective action, etc.).

It should be noted that CENELEC 50128:2011 only refers to ISO 9001:2008, but a company may have a quality assurance mechanism which conforms to a different framework such as Capability Maturity Model for integration (CMMi[4]) or Software Process Improvement and Capability

3 "Replay" is the ability to repeat activities precisely ("doing and redoing" something). For effective replay, it is important that nothing should be left implicit.

4 CMMi is a maturity model specific to the software industry which encapsulates a set of good practices to be employed in development projects. For further developments, see the website of the Software Engineering Institute: www.sei.cmu.edu/cmmi.

dEtermination ((SPICE) – see [ISO 04b]), but in that case, we must provide evidence (an equivalence matrix such as Table 4.2) to demonstrate the equivalence with ISO 9001:2008 [ISO 08b].

In the context of railways, IRIS[5] [UNI 09] has been defined to integrate the respect of ISO 9001:2008 and of the CENELEC framework (50126, 50128 and 50129) into a single certification.

4.3.2.3. *Indicator*

ISO 9001:2008 recommends the establishment of indicators by which to monitor the quality of the software product and of the process in place.

The 2001 version of CENELEC 50128 [CEN 01a], in section 15.4.5 – item 7 – called for the definition of metrics (in the sense of quantitative measures) to be applied to the process and the product. It introduced an explicit link to ISO/IEC 9126 [ISO 91, ISO 04c].

ISO/IEC 9126 defines and describes a series of quality characters of a software product (internal and external characteristics, usage features). These characteristics can be used to specify the functional and non-functional requirements of the customers and the users.

The characteristics identified by ISO/IEC 9126 are:

– functional capacity: does the software cater for the functional requirements expressed?

– reliability: can the software maintain its level of service in precise conditions and for a specific length of time?

– usability: does the software require little effort to use?

– efficacy: does the software require profitable, proportionate sizing of the hosting platform in relation to the other requirements?

– maintainability: does the software require little effort in order to evolve to cope with new requirements?

– portability: can the software be transferred from one platform or environment to another?

5 "IRIS" stands for International Railway Industry Standard. For further information, visit: www.iris-rail.org.

CENELEC 50128:2011 [CEN 11a] and 62279:2014 [IEC 14] have transferred the demand for respect of ISO/IEC 9127 to the level of realization of the software specification requirements (section 7.2.4.1 of CENELEC 50128).

CENELEC 50128:2011 introduces the need to master the complexity as illustrated by row 8 of Table A.12. This command of the complexity is "HR" for all SSILs.

In order to take account of the state of the art, it would have been preferable to take account of ISO 25000 [ISO 14] for the requirements and evaluation of software quality. That standard is also known as "SQuaRE", which stands for Software QUAlity Requirements and Evaluation.

4.3.3. *Realization of a software application*

It should be noted that we speak of the "realization" rather than the "development" of a software application. The realization of a software application comprises the activities of development but also those of verification, validation, production, installation and maintenance of the software application (see Figure 4.1).

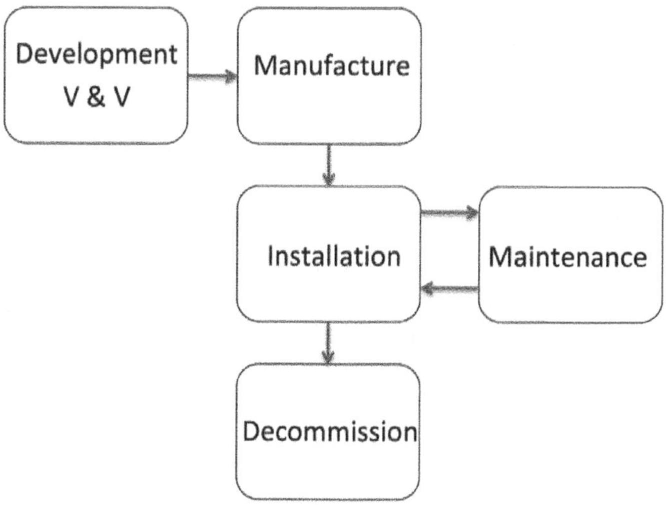

Figure 4.1. *Steps in the realization of a software program*

The activities of verification of validation are important, and will be carried out more or less fully depending on the level of safety required. As regards the activities of production of the final application and installation, they are crucial, and require specific processes to be put in place. The decommissioning (or "retirement") or a software application is mentioned, but poses no major problem, unlike the retirement of a complex system, such as the decommissioning of a nuclear plant or a railway installation.

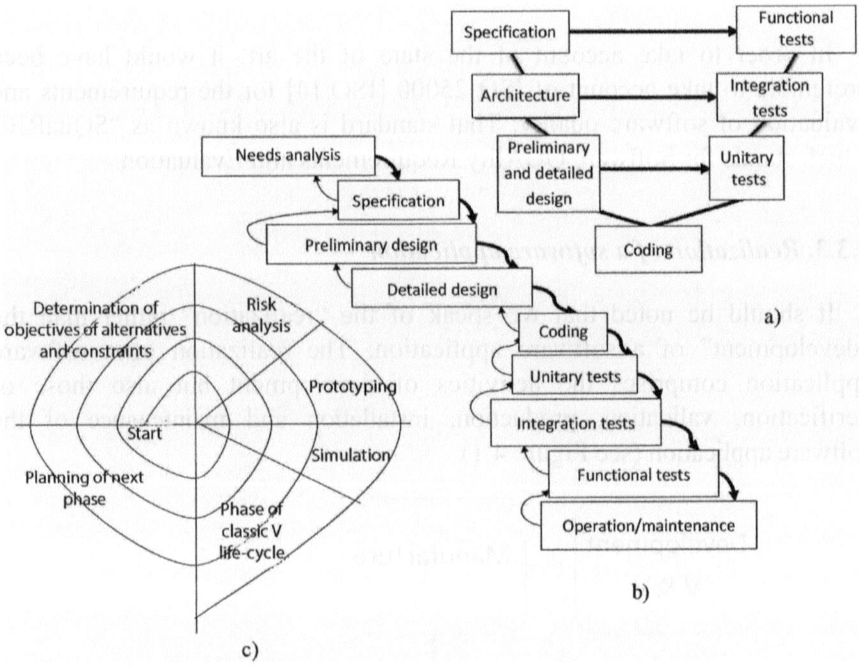

Figure 4.2. *Different cycles in the realization of a software program*

The maintenance of the software application remains the trickiest activity. Indeed, in the wake of a change, it is crucial to maintain a level of safety whilst keeping the cost of the evolution under control and minimizing the impact on a system in service.

There is a difficulty facing the maintenance of a software application: the lifespan of that software application. Indeed, for the railway domain, the lifespan is between 40 and 50 years; in aeronautics it is forty years; in the nuclear industry, 50 years; and for the automotive industry, 15 years. In light of these expected lifespans, it is necessary to take measures to ensure continuity of service and maintenance of the software application.

As Figure 4.2 illustrates, there are various cycles ((a) V cycle, (b) cascade cycle, (c) spiral cycle, etc.) for the realization of a generic software application, but the cycle recommended by the various standards (CENELEC 50128:2011, IEC 61508, ISO 26262) is still the V cycle.

Finally, we need to define a cycle of realization, and to stay on top of different aspects such as: application of the realization cycle, requirements traceability management, change management, configuration management, tool management, etc.

4.3.4. *Software quality assurance plan (SQAP)*

Thus, for any given project, we must identify the process of realization of the software (e.g. the V-cycle shown in Figure 4.3), the quality objectives, the procedures employed to realize the product and the realization conditions. Today, it is the V-cycle that is recommended by the various standards applicable to the realization of safety-related software.

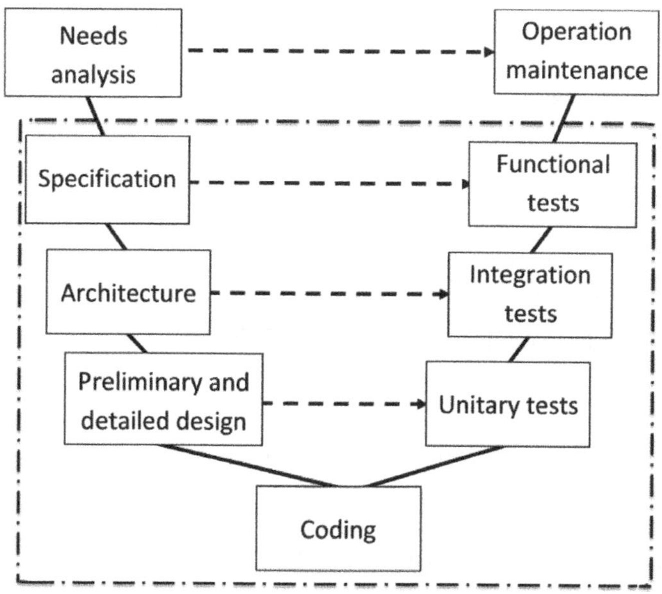

Figure 4.3. *V-cycle*

With this in mind, a Project Quality Plan (PQP) must be drawn up for each project. The PQP is a document describing the specific techniques employed by a company to ensure the quality of the product or service to which the project pertains. As a general rule, not all of the procedures implemented by a company will be applicable to the project, and certain specific aspects of that project will need to be taken into account (e.g. innovation, outsourcing of activity, etc.).

Quality management for the software application involves the implementation of a "Software Quality Assurance Plan" (SQAP). Indeed, to ensure quality during the realization of a software application, one must define the resources that need to be mobilized.

By "resources", we mean human resources, material resources, and resources in terms of methods, processes and tools. There are a number of objectives attached to the SQAP (see Appendix A):

– to define the organization of the project (for instance, see the form of organization identified in Figure 4.6);

– to demonstrate the suitability of the assigned personnel for their respective roles;

– to define the software realization cycle, which must take account of the aspects of development, verification, validation and assessment. CENELEC 50128:2011 advocates the use of a V-cycle (similar to that illustrated by Figure 4.3), but it is possible to define other realization cycles;

– to identify the procedures which are applicable, the applicable guides, the input elements, output elements and activities to be performed (see Figure 4.4) at each phase of the cycle;

– to demonstrate conformance to CENELEC 50128:2011;

– etc.

In view of the organization of the project, as a general rule, the SQAP introduces the lifecycle and fully defines the design phases, involving a Software Configuration Management Plan (SCMP), a Software Verification Plan (SVeP) and a Software Validation Plan (SVaP).

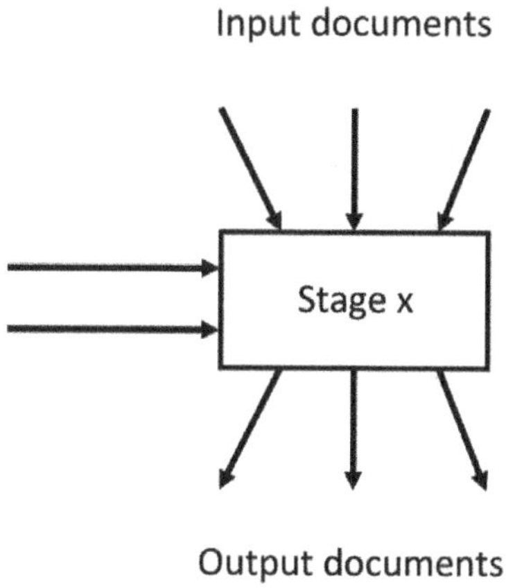

Figure 4.4. *Description of a step*

In certain cases, the SVeP and SVaP are combined in a Verification and Validation Plan (VVP), which covers the verification of each phase of the lifecycle, the unit tests, the integration tests (examining software/software and hardware/software integration) and the target validation tests.

The plans pertaining to the software realization (SQAP, VVP and/or SVaP, SVeP) need to be verified, and the Quality Verification Report (QVR) must contain the results of that verification. This verification of the plans should demonstrate that the company's QAM and the quality objectives for the target SSIL have been respected.

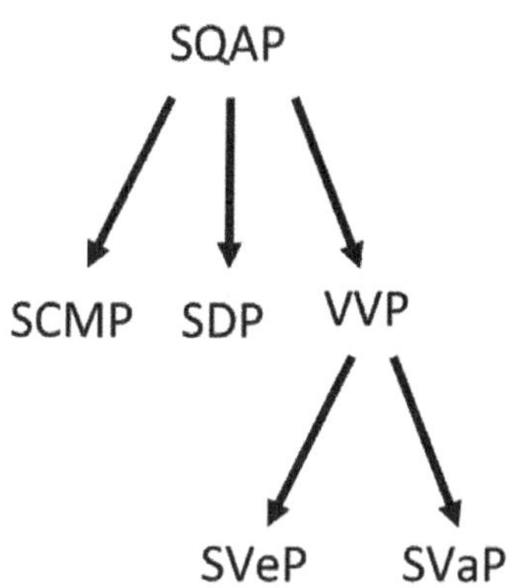

Figure 4.5. *Hierarchy of plans*

It is necessary to ensure the quality of the project; in order to do so, we must put in place a set of metrics by which to evaluate the project and identify any difficulties. The possible metrics include: number of errors, number of versions, number of requirements dealt with and/or tested, etc. These metrics are essential, as they enable us to identify potentially-problematic situations, such as a peak in demand, improper quality management, lack of skill, etc.

4.4. Organization

4.4.1. *Typical organization*

CENELEC 50128:2011 suggests three types of organization, which depend on the SSIL (SSIL0, SSIL2 and SSIL3/4). Figure 4.6 represents a complete organization for a software realization project of SSIL3/4.

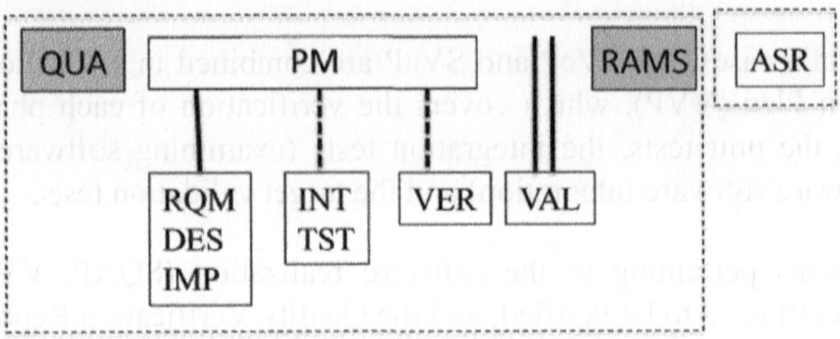

Figure 4.6. *Typical architecture for an SSIL3/4 project as described by CENELEC 50128*

The shaded boxes in the above figure contain two roles (QUA and RAMS) which are not in the CENELEC 50128:2011 framework but are introduced by the application of ISO 9001:2008 and CENELEC 50126 [CEN 99].

From Figures 4.6 and 4.7, we are able to identify the hierarchical links and the levels of independence. The dotted-line boxes represent two independent organizations.

The realization of a program with SSIL4 will involve at least eight people on the project, who will need to assume 11 roles:

– QUA: the quality engineer is in charge of verifying the proper implementation of the quality procedures and the company's rules in the project. He/she must also verify the correct application of the plans (SQAP, VVP, SCMP, etc.);

– RAMS: the safety engineer is in charge of managing the safety analysis related to the desired SSIL for the software;

– PM: the project manager is in charge of managing and organizing the activities involved in the realization;

– RQM: the requirements manager is responsible for clearly laying out the requirements (see Table 4.2);

– DES: the designer is responsible for constructing the architecture and coming up with the design for the software;

– IMP: the implementer is charged with producing executable code on the basis of the design;

– INT: the integrator is in charge of the integration of software components that have already been tested. This integration continues until a complete software package has been constructed;

– TST: the tester's task is to perform the component test and the overall software test;

– VER: the verifier's task is to perform verifications and checks, which may relate to a document, a file, a process, etc.;

– VAL: the validator must, on the basis of the various activities, confirm whether or not the software can be considered valid;

– ASR: the assessor is independent of the realization team, and must reach a verdict as to the achievement of the SSIL.

Appendix B of CENELEC 50128:2011 introduces a description of the responsibilities and skills required for each role. This breakdown makes it clear that no one person can correctly realize a software program: a diverse set of skills is required for each phase.

It should be noted that in IEC 62279:2014 [IEC 14], quality was taken into account by the introduction of the role of the Quality Assurance Manager (QUAM) added alongside the VER. IEC 62279:2014 thus advocates the same approach, but new roles have been introduced (see Figure 4.7), such as the Configuration Manager (CGM) and the *reviewer* (REV), and therefore Appendix B in that document is fuller:

– QUAM: the quality assurance manager is the person in charge of the SQAP and of verifying that the plans are correctly implemented;

– CGM: the configuration manager is the person in charge of overseeing the configuration and producing the associated documents (SCMP, SwVS[6]);

– REV: the reviewer is the person in charge of carrying out a review of the documents output at each phase of the project.

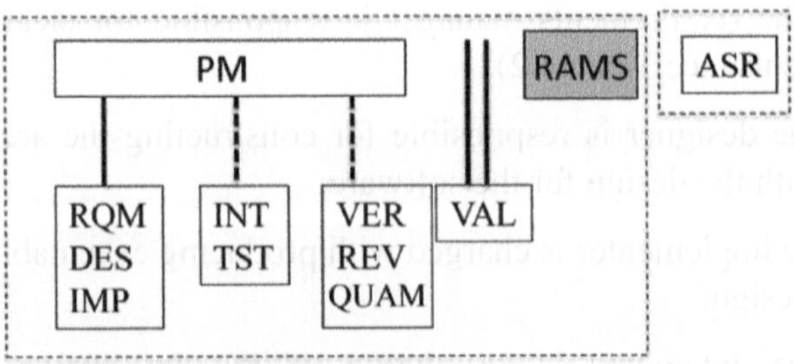

Figure 4.7. *Typical architecture for an SSIL3/4 project as described by IEC 62279*

4.4.2. *Skill management*

Clause 5 of CENELEC 50128:2011 places emphasis on the management of the skills at work during the project and on the formalization of that management process. Normally, the SQAP must contain elements demonstrating the skills of the people involved for the roles that they assume, but this management may be formalized at company level.

This skill management and/or justification cannot be based solely on the management of curricula vitarum (CVs). It is crucial to be able to properly demonstrate that the personnel are adequately qualified for their various roles.

One of the difficulties in skill management lies in the management of the skills of people outside the company (at subsidiary companies, subcontractors, experts to whom work is outsourced, etc.). The labor code and the commercial legislation include the prohibition to manage the skills of external workers, but ISO 9001:2008 does not allow for external skill management.

6 SwVS for Software Version Sheet.

In railway projects, it is important to put in place a managerial strategy (job description, skill sets, identification of requisite prior training, etc.), which can be used to demonstrate that the external personnel are adequately skilled and are capable of assuming the roles assigned to them.

For subcontracting (where work is done for a flat fee by an outside party), it is necessary to demonstrate that the subcontractor has a skill-management scheme and is able to assign the various roles in the projects to the appropriate people, but it is crucial for the customer to perform at least one audit to check that this skill management works properly. In general, this audit of the subcontractor is performed at the same time as the audit to demonstrate ISO 9001:2008 conformance.

Appendix B in CENELEC 50128:2011 and in IEC 62279:2014 presents the roles and, for each role, identifies the responsibilities and skills required (see Table 4.3). It should be noted that for each role, we find the need for knowledge (at least familiarity with the part of the project in question) and for familiarity with the regulatory framework.

Role: requirements manager: RQM
Responsibilities: – must be in charge of specifying the software requirements; – must deliver the specifications laid out by the software requirements; – must establish and maintain traceability to and from the requirements at system level; – must ensure that the requirements relating to the specifications and the software are taken into account in the management of the modifications and of the configuration, including the state, version and authorization status; – must ensure consistency and completeness in the specification of the software requirements (with reference to the requirements of the end user and the final environment of application); – must develop and maintain the requirement documents relating to the software.
Main skills: – must be competent in requirements engineering; – must have experience in the field of application; – must have experience in the safety criteria in the field of application; – must understand the overall role of the system and the environment of its application; – must understand the analytical techniques and their results; – must understand the applicable regulations; – must understand the requirements set out by CENELEC 50128.

Table 4.3. *Extract from Table B.1 in CENELEC 50128:2011*

For management of the external people involved in the project, it is important to request certification of training in CENELEC 50128:2011 (the request may also be extended to the whole of the CENELEC framework) and in the legislative context.

Figure 4.8 introduces an example of a process of resources which is organized around a phase of allocation. At entry into the allocation phase, we have job sheets which, for each role, define the requisite skills and level of training, and the list of people both within and outside the company who need to be involved in the project. Upon completion of the allocation phase, we must find justifying factors which demonstrate that the people to whom each role has been assigned are competent for those roles. In order to do so, we need to be capable of linking a person to a role, and we must have elements of justification which show that the requisite skills have been acquired, or that further training must be undertaken.

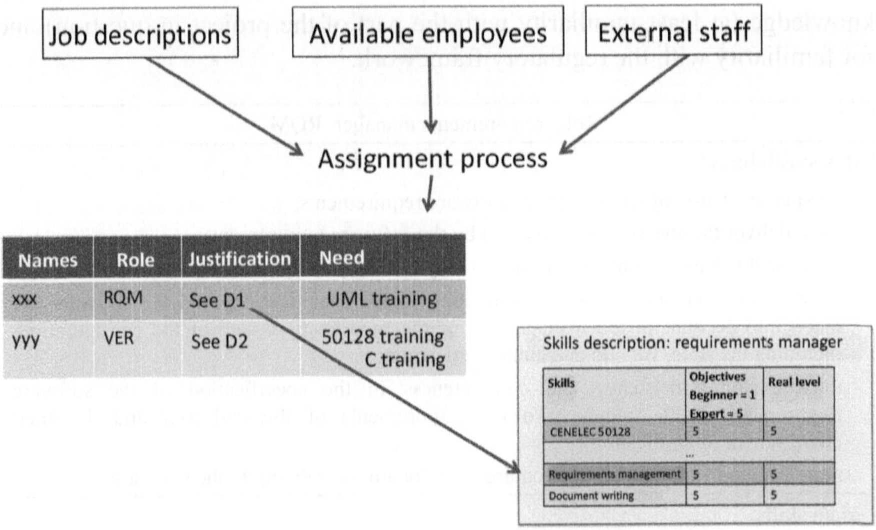

Figure 4.8. *Process of resource and skill management*

4.5. Configuration management

The software forms part of a set of a piece of equipment whose configuration must be managed. For this reason, at system level, we must have a "Configuration Management Plan" (CMP). The CMP must cover all aspects of the equipment: pneumatic, mechanical, electrical, electronic, hardware, software, maintenance tools, etc. The CMP describes the naming conventions, the rules of version management, the means of configuration management, the organization, the responsibilities, etc., and the contents of the system version sheet (called SysVS).

With regard to software configuration management, we must be capable, at all times, of pinpointing the list of elements produced during the realization of the software application and the associated versions.

It should be pointed out that the sources and the process of generation of the executable are only a few elements stemming from the realization of the software, and that it is important not to forget all the documents produced (plans, specifications, design documents, test files, verification documents, end-of-phase report, etc.), the various scenarios and test results[7] (CTs, H/S and S/S ITs, OSTs), the results of the tooled verification phases (coverage synthesis file, metrics synthesis file, code analysts' report, etc.) and the tools implemented throughout the process.

Role: configuration manager: CGM
Responsibilities: – must be party to the configuration management system; – must establish that all the requirements relative to the software are clearly identified and attached to the appropriate versions independently within the configuration management system; – must prepare the delivery sheets which mention incompatible versions of the software components.
Main skills: – must be competent in software configuration management; – must have an understanding of the requirements set out by IEC 62279.

Table 4.4. *Extract from Table B.10 in IEC 62279:2014*

Successful software configuration management involves the establishment of the software configuration management plan (SCMP). The objective of the SCMP is to define the general approach (what, when and how), the tree structure(s) to archive all of the elements (source, documents,

7 The strategies of component tests (CTs), integration tests (ITs) and overall software tests (OSTs) will be presented in Chapter 6.

files, tools, etc.) and the people in charge of managing and controlling the configuration management. The proposed process must define the content of the software version sheet (SwVS). IEC 62279:2014 introduces a specific role for this activity and a job specification (see Table B.10).

The configuration management plan must identify a control activity: it is not acceptable to discover at the end of a project that elements have been lost or that the configuration is inconsistent. This activity of verification can be carried out by means of configuration audits. These audits can be performed by the quality team.

4.6. Safety assurance management

CENELEC 50128:2011 does not identify a safety assurance activity, but it requires the implementation of CENELEC 50126. CENELEC 50126 defines a cycle for management of the FMDS (see Figure 3.4), which is based on the acquisition of the needs when the system is removed. For our purposes though CENELEC 50126 requires the safety team to identify the safety-related functional requirements and the safety-integrity requirements, and that they follow all of the requirements throughout the whole of the cycle.

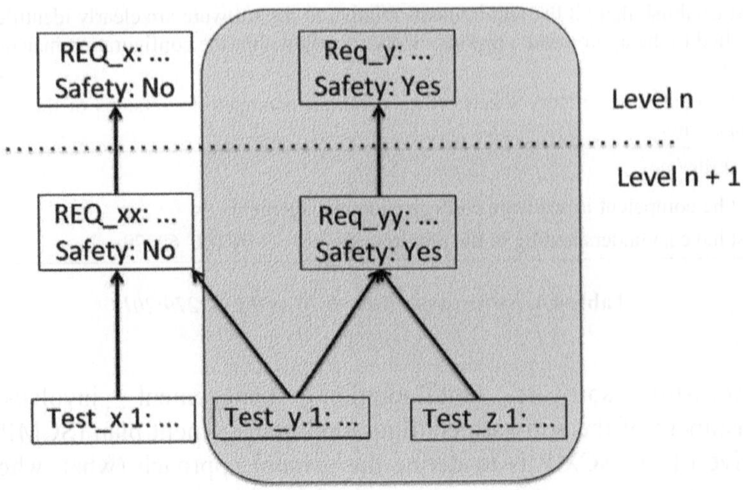

Figure 4.9. *Example of requirements traceability*

For this reason, the safety team must verify that all the safety-related requirements are served by the software application at the stage of its realization, but also from the moment of its entry into service to its ultimate decommissioning. The safety team therefore must verify that each safety-related requirement is correctly dealt with during the realization of the software. Hence, they must analyze all of the residual faults in the software in order to verify the impact on the system, and analyze every change made during the maintenance phase in order to demonstrate that these changes do not impact on safety.

With regard to the verification of the requirements, Figure 4.9 presents an example of a situation. The upper-level requirements are broken down at the following levels and associated with tests. The responsibility of the safety team is to verify that the safety requirements are correctly dealt with, which is modeled by the gray box.

The activity of the safety team in the context of a software program must hinge on the following activities:

– identification of the safety requirements applicable to the software, by synthesis of the safety documents such as the "Preliminary Hazard Analysis" (PHA), "System Hazard Analysis" (SHA), "Interface Hazard Analysis" (IHA), "Sub-System Hazard Analysis" (SSHA) and "Hazard Log" (HL);

– participation as checker in all document reviews, to demonstrate that the safety-related requirements are dealt with by the software;

– analysis of coverage of the safety requirements by the tests (unit tests, component tests, integration tests, full system tests);

– analysis of the event reports describing the residual errors;

– identification of the exported constraints on the use of the software;

– *a priori* analysis of the demands for evolution of the software application;

– specific analyses linked to the demonstration of the software safety, such as software error effects analysis (SEEA – see [AFN 90] and [GAR 94]) and/or critical code reading (CCR);

– independent assessment management: it is necessary to make the connection with the independent safety assessor (ISA) and to manage the impact of comments and questions on the project.

Figure 4.10 presents the safety management cycle as prescribed by CENELEC 50126, but positioning a few activities.

CENELEC 50126 recommends that the work of the safety team be described in the safety assurance plan (SAP), but in view of the specificity of the activities linked to software safety demonstration, it is preferable to establish a software safety assurance plan (SSAP) which focuses on the software.

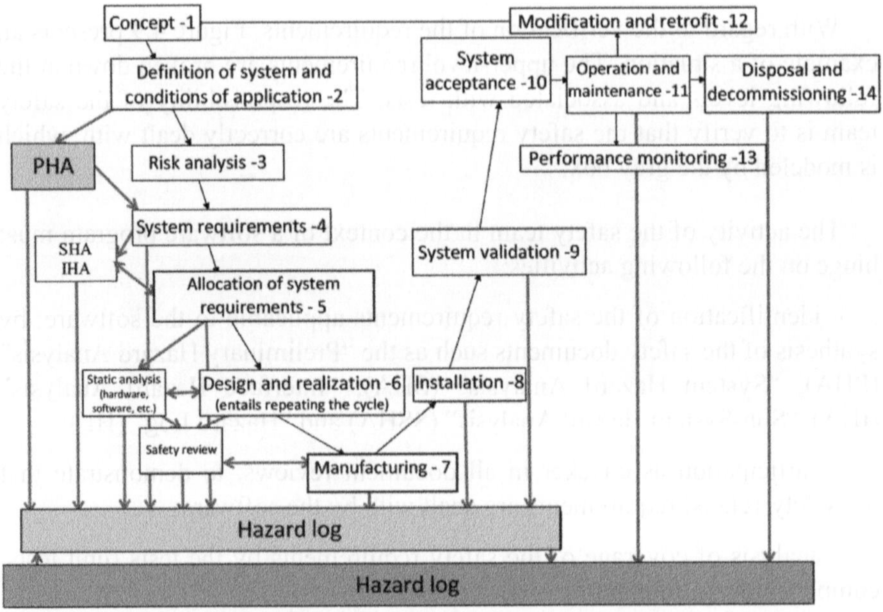

Figure 4.10. *Annotated safety cycle as prescribed by CENELEC 50126*

4.7. Verification and validation

4.7.1. *Introduction*

The realization of a software application must take account of the design of that application, but it must also take account of the activities whereby it is possible to demonstrate that the application achieves a certain level of quality. The attainment of a level of quality involves demonstrating that no errors have been introduced during the design phase and that the product corresponds to the requirements which were initially identified.

As regards V&V management, the activities must be defined and positioned at an appropriate point in the lifecycle of the software's realization (see Figure 4.8). It is important to specify the organization and responsibilities of the V&V team, and thus draft one or more plans (see Figure 4.5).

In view of the rules of independence introduced by CENELEC 50128:2011, the people performing the verification may or may not be in the same team as those in charge of validation. The independence requirements are linked to the SSIL objective:

– for an SSIL0, the PM can manage the whole of the realization team but not the quality, safety and evaluation teams. The project team must be divided into two parts: a development part and a V&V part;

– for an SSIL2, the PM can be in charge of the realization team but not of the quality, safety and evaluation teams. The project team must be divided into three parts: a development part, a test realization part and a V&V part;

– for a SSIL3-SSIL4, in addition to the above restrictions, the PM cannot be in charge of the validation team (see Figures 4.6 and 4.7).

4.7.2. *Verification*

4.7.2.1. *Presentation*

Figure 4.11 illustrates the main problem of the realization of a software application. Indeed, there is a need to realize and at the same time there is a realization. The purpose of verification is to demonstrate that all the need is taken into account by the realization and that no unexpected elements exist. The development team will always have good reasons for introducing undesired code fragments (recycled functions, addition of a service, etc.) and for not catering for all the requirements (technical difficulties, oversights, etc.).

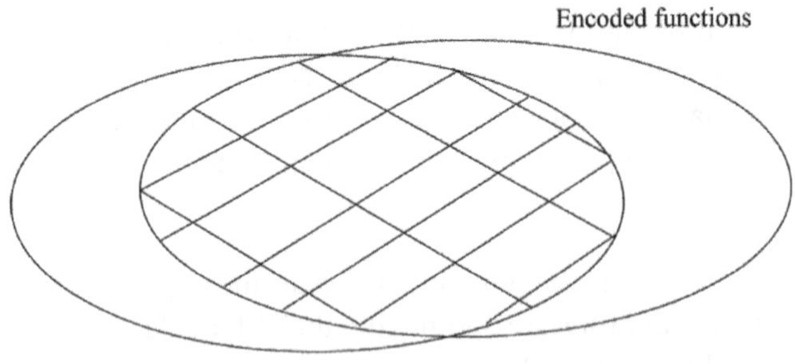

Figure 4.11. *Verification and validation*

ISO 9001:2008 recommends that each production be systematically verified. CENELEC 50128:2011 stresses the point, indicating that the role of the verification is to show that the product has been correctly created (i.e. that no faults have been introduced). Thus, as Figure 4.12 illustrates, the verification is an activity which covers every phase of the lifecycle.

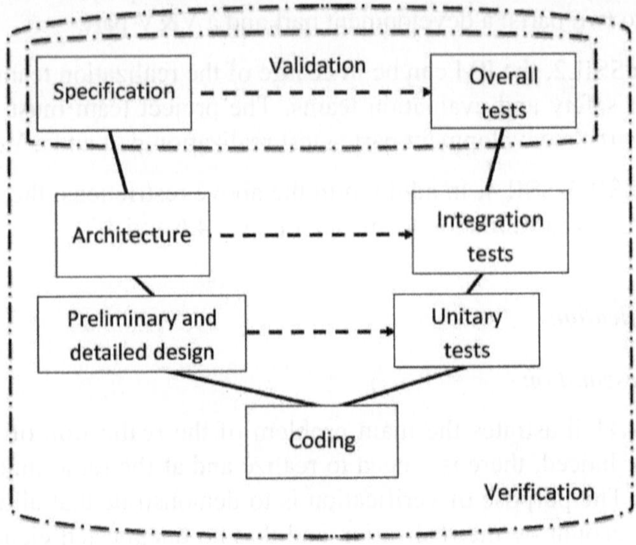

Figure 4.12. *Verification[8] and validation*

8 In the figure, we have introduced the unit tests, but it is also possible to see module tests or component tests.

As Figure 4.13 shows, on the basis of the inputs and the activities, we need to demonstrate that the output elements are correct.

In addition to the elements discussed previously, we must indicate the need to have elements of proof which are able to demonstrate that the activity of verification has been correctly performed and that the product respects the input requirements – hence Definition 4.2.

DEFINITION 4.2 (Verification).– *Confirmation by tangible proof that the specified requirements have been satisfied at each stage in the realization process.*

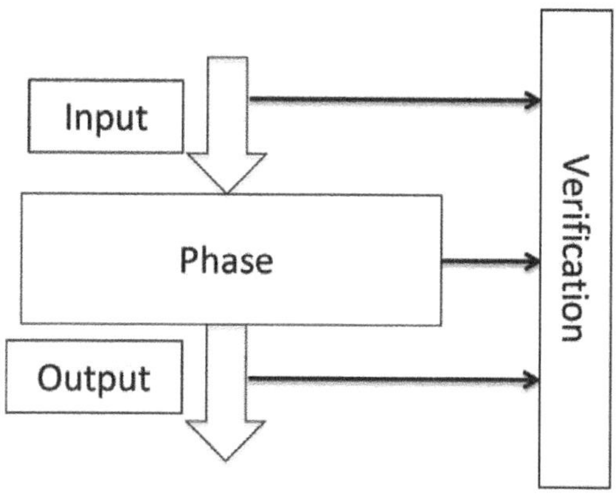

Figure 4.13. *Verification*

The verification of a phase requires us to analyze the implementation of the quality requirements (application of the procedures, respect of formats, etc.), the application of the processes (respect of plans, respect of the organization, etc.), the correctness of the activities and the accurate integration of the safety requirements.

In summary, the objective of verification is to show that we have correctly made the product. Thus, the associated question is: has the work been well done?

4.7.2.2. *Activity of verification*

In the context of CENELEC 50128:2011, verification is the activity which consists, for each phase of the lifecycle, of demonstrating by analysis and/or testing that, for the specific inputs, at every point the deliverable elements fulfill the objectives (identified in the plans) and the requirements set for the phase.

Below is a non-exhaustive list of the activities of verification:

– the reviews relating to the outputs from a phase are designed to ensure conformity with the objectives and prescriptions for that phase, taking account of the inputs specific to that phase;

– design reviews: code analysis, verification of coding rules, etc.;

– the whole-program tests performed on the product in order to ensure that the function conforms to the specification;

– the integration tests performed during the element-by-element assembly of different parts of a system, based on environmental tests, in order to ensure that all the parts function with one another in line with the specifications;

– the component test performed after the realization of a component. At that point of the presentation, we find the link between the component test and the module tests and/or unit tests;

– etc.

Section 6.2 of the standard presents the activity of verification. This activity must be described (in the VVP or in an SVeP), and it must be justified (see sections 4.8, 4.9 and 4.10 of the standard). The plan (VVP or SVeP) must be verified, and the QVR must contain the results of that verification. This verification of the plan must be capable of showing that the company's QAM has been respected and that the objectives set out in CENELEC 50128 for the SSIL objective are met.

As the verifications are performed for each phase, it is necessary to produce a quality verification report for each of those phases. The production of that QVR may be coupled with an end-of-phase review.

Measure	SSIL0	SSIL1 SSIL2	SSIL3 SSIL4
1. Formal proof	–	R	HR
2. Static analysis (A.19)	–	HR	HR
3. Dynamic analysis and dynamic tests (A.13)	–	HR	HR
4. Metrics	–	R	R
5. Traceability	R	HR	M
6. Software Error Effects Analysis	–	R	HR
7. Coverage of test for the code	R	HR	HR
8. Functional tests and black-box tests	HR	HR	M
9. Performance tests	–	HR	HR
10. Interface tests	HR	HR	HR

Table 4.5. *Extract from Table A.5 in CENELEC 50128:2011*

Section 6.2 is based on Table A.5 (reproduced above as Table 4.5) which identifies two "families" of activities:

– static verifications (which do not require execution): rows 1, 2, 4, 5 and 6;

– dynamic verifications (which require execution): rows 3, 7, 8, 9 and 10.

4.7.2.3. *Static analysis*

In this section, we shall focus on static analyses. Static analyses are a way of performing a check without executing the code of the application.

Measure	SSIL0	SSIL1 SSIL2	SSIL3 SSIL4
1. Boundary-value analysis	–	R	HR
2. Control lists	–	R	R
3. Control flow analysis	–	HR	HR
4. Data flow analysis	–	HR	HR
5. Error guessing		R	R
6. Design reviews/checks	HR	HR	HR

Table 4.6. *Extract from Table A.19 in EN 50128:2011*

In Table A.5, traceability (for instance, see Table 4.7) and the metrics are reiterated as a means of verification in addition to Table A.9 (Table 4.1), with the difference being that the metrics are now merely "R" (as opposed to "HR") for verification. Indeed, the verification cannot solely be based on metrics.

Software Architecture File (DAL from French)	Preliminary Design File (DCP from French)
DAL_EX_1	DCP_EX_11, DCP_EX_12, DCP_EX_13
DAL_EX_2	
DAL_EX_3	DCP_EX_11

Table 4.7. *Example of traceability matrix between DAL and DCP*

```
void main()
{
    int x = 0;
    int y = 1;
    while (y < 10)
    {
        y = 2 * y;
        x = x + 1;
    }
    printf ("%d",x);
    printf ("%d",y);
}
```

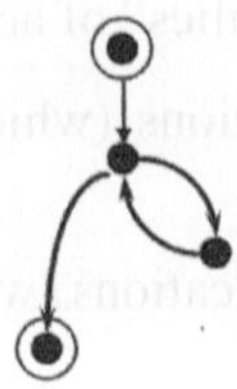

Figure 4.14. *Example of a control graph*

Below are a few metrics (for examples, see [CAB 96]):

– cyclomatic number v(G): the cyclomatic number describes the complexity of a program. It quantifies the number of paths of execution of a procedure/function. This metric is measured from the control graph (see Figure 4.14);

– number of lines of code: there are several variants of this metric (with or without comments, with or without blank lines). This metric can be used to evaluate readability and maintainability (presence of comments);

– Halstead metrics: this array of metrics can be used to evaluate the complexity of a program fragment by examining the complexity of the expressions used (the operators and operands);

– number of errors detected, number of errors corrected, number of evolutions.

The cyclomatic number is one of the most prevalent and important metrics, because it enables us to establish a minimum limit for the number of unit tests that need to be performed to test a code fragment.

It is important to note that it is possible to construct direct metrics (relating to the code) or indirect metrics (synthesis of other metrics, process analysis, etc.).

By way of example, Table 4.8 shows an example of the results of measures taken by metrics on a code using the analyzer "Logiscope". The table shows the name of the metric, the minimum and maximum limits (chosen for the project or for the company) and the measure which must be employed.

Metric	Name	Min. value	Max. value	Value
Number of commands	N_STMTS	1	50	70
Program length	PR_LGTH	3	350	85
Cyclomatic number	VG	1	20	11
Max number of levels	MAX_LVLS	1	5	3
Number of non-cyclical paths	N_PATHS	1	80	29
Number of "Goto" jumps	N_JUMPS	0	0	0
Comment rate (frequency)	COM_R	0.2	1.0	0.17
Average length of commands	AVG_S	3	7	1.21
Number of input and output points	IO_PTS	2	2	2

Table 4.8. *Example of a table containing the result of measures gained by various metrics*

As regards the applicability of measure theory in object-oriented languages, there is no answer that can be given at present. Many different metrics have been developed (see [CHI 94]), but they hinge on the aspect of complexity of the structure, so it is difficult to establish a clear relationship between them and the conventional indicators of quality (testing effort required, maintainability, etc.).

In terms of static analysis, there is a link to Table A.19 (see Table 4.6). Table A.19 cites control lists as "Recommended", because they cannot be employed alone for a verification; rather, they represent a guide which identifies objectives.

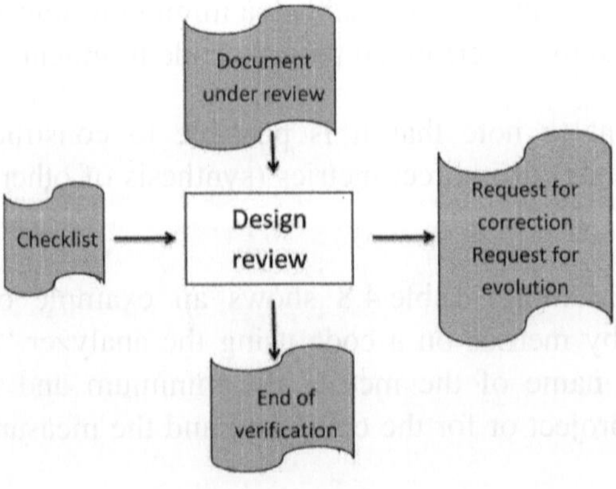

Figure 4.15. *Design review*

From Tables A.9, A.5 and A.19, we must deduce that the use of reviews (see Figure 4.15) is highly recommended (HR) – an obligation, even – and that in order to perform these reviews it is necessary to:

– identify the complex elements by using metrics;

– have control lists (for an example, see Table 4.9) to guide the checking process and set clear verification objectives. Control lists help to identify the elements of justification which need to be produced;

– verify the validity of the traceability. Traceability is an input and an integral part of the documents produced during the process of realization of the software. One of the verification team's goals is to verify the traceability;

– perform flow analyses (control flow/data flow). Using flow analyses, it is possible to verify that any variables to be consumed have been produced, that all produced variables are consumed, that the architecture of the call diagram is correct, etc.;

– have a formalized review process.

Point	Issue	Status: OK/KO	Comment
R_1	Is the document title correct?		
R_2	Is the document reference correct?		
R_3	Has the version number been specified?		
R_4	Are the title, reference and version number present on each page?		
R_5	...		
R_6	Is the list of input documents given?		
R_7	Is it possible to identify the versions used?		
R_8	...		

Table 4.9. *Example of a checklist*

Checklists (for example, see Table 4.9) must define checkpoints. These checkpoints are linked to knowledge of the types of errors which may be introduced during the activity having led to the production of the documents being verified.

Row 6 of Table A.19 also mentions design analysis, which refers to verification that the various guides (modeling guide, design guide, coding guide, etc.) have been taken into account. Indeed, Tables A.4 and A.12 in CENELEC 50128:2011 identify the need for a guide, so it is important to check that these guides have been correctly followed.

Formal proof is cited as a means of verification that can be used. The use of formal proof can replace the testing activities, but if so, it is necessary to explain and justify how the proof is equivalent to the corresponding battery of tests. In any case, such proof cannot replace the whole-program testing activities (functional tests) using the software on its target platform.

4.7.2.4. *Dynamic analysis*

4.7.2.4.1. Test strategy

Table A.5 illustrates the need to perform functional tests (L8) and interface tests (L10) from SSIL0 onwards.

Measure	SSIL0	SSIL1; SSIL2	SSIL3; SSIL4
1. Execution of a test dossier on the basis of boundary-value analysis	–	HR	HR
...			
4. Modeling of performances	–	R	HR
5. Equivalence-class tests and input-partitioning	R	R	HR
6. Structural tests	–	R	HR

Table 4.10. *Extract from Table A.13 in CENELEC 50128:2011*

The conduction of tests remains the principal activity, which is performed in three steps:

– a step of selection of test cases (Table A.13-L1), which results in the compilation of a test dossier (DTx where x = C, I or L);

– a step of encoding of the test cases in computing scenarios;

– a step of execution of the computing scenarios and formalization of the results in a dossier of test results (DRTx where x = C, I or L).

Measure	SSIL0	SSIL1; SSIL2	SSIL3; SSIL4
1. Functional tests and black-box tests	HR	HR	HR
2. Performance tests	–	R	HR

Table 4.11. *Extract from Table A.6 in CENELEC 50128:2011*

One of the important points is the selection of test cases, which must respect three criteria:

– black-box tests (A.5-L8, A.6-L1, A.7-L2);

– boundary-values analysis (A.13-L1, A.14-L3);

– selection of test cases after identification of equivalence classes (A.13-L5, A.14-L4).

As regards black-box tests, this is the most important point. Constructing black-box tests consists of not having design elements to identify test cases. Therefore, during the design phase, the code is not accessible to the team in charge of identification of the test cases, and therefore it is necessary for the design elements (detailed design dossiers, for example) to be sufficiently detailed (presentation of the requirements, of the algorithms and of all the data being handled). Similarly, it is not possible to take the formal model which served to generate the code as input to the whole-program testing and/or integration testing phase.

Table A.7 in CENELEC 50128:2011 relates to the whole-program tests and introduces the possibility of creating a model of the requirements of the software specification to selection the test cases (see [BOU 99] for an example of the implementation).

Measure	SSIL0	SSIL1; SSIL2	SSIL3; SSIL4
1. Performance tests	–	HR	M
2. Functional and black-box tests	HR	HR	M
3. Modeling	–	R	R

Table 4.12. *Extract from Table A.7 in EN 50128:2011*

CENELEC 50128:2011 identifies the need to carry out performance tests and leaves open the possibility of carrying them out during the integration phase and/or the phase of testing of the whole of the software. These tests are extremely important and necessitate the identification of the performance requirements (A.13-L4, A.18): response time, cycle time, processor workload, memory workload, etc.

Measure	SSIL0	SSIL1; SSIL2	SSIL3; SSIL4
…			
3. Boundary-value analysis	R	HR	HR
4. Equivalence classes and input partitioning tests	R	HR	HR
…			

Table 4.13. *Extract from Table A.14 in CENELEC 50128:2011*

We shall specify in the description of the downward phase in the V cycle (Chapter 7) how this section must be applied for the different testing activities: component tests, interface tests and tests of the whole software package.

4.7.2.4.2. Coverage of tests

One of the important issues for a project is: “How do we decide when to stop the tests?”. There are a variety of possible responses:

– a sufficient level of confidence has been attained (purely subjective);

– an objective of coverage of the instructions has been achieved;

– structural coverage has been achieved;

– an objective of coverage of the inputs and outputs has been achieved;

– coverage has been achieved;

– the estimated number of residual errors is acceptable;

– etc.;

– and as the final possibility, the budget has run out.

In fact, for critical applications which have an impact on safety, the coverage does not help the decision as to when to terminate the tests: it has another, more important objective, which is to identify the portions of code not covered (i.e. which are not stimulated, or not executed) by the tests. As is indicated by item 1 on the requirements given in Table A.21, the purpose of coverage is to increase confidence in the activity of testing. Any portions of code not covered are linked to one of the following situations:

– case 1: the tests performed are incomplete in relation to the specifications, and it is then necessary to supplement the tests but also to understand why the tests have not been specified;

– case 2: the code of the application contains code which is not related to the specifications (i.e. additional codes alongside the specifications).

Measure	SSIL0	SSIL1; SSIL2	SSIL3; SSIL4
1. Instruction	R	HR	HR
2. Branch	–	R	HR
3. Compound conditions	–	R	HR
4. Data flow	–	R	HR
5. Path	–	R	HR

Table 4.14. *Extract from Table A.21 in CENELEC 50128:2011*

In fact, to be more specific, case 1 can be linked to the fact that it is not possible to perform any tests (defensive programming, bad code structure, etc.) or that genuine oversights have occurred in the identification of the tests. For case 2, either the coder has added code which is not linked to the specifications, or he/she has applied programming rules (e.g. defensive programming) which introduce cases that are unidentifiable on the basis of the specifications.

The 2001 version of CENELEC 50128 called for a demonstration of the coverage of the instructions. In its 2011 version, 50128 introduces a more specific requirement in its Table A.21.

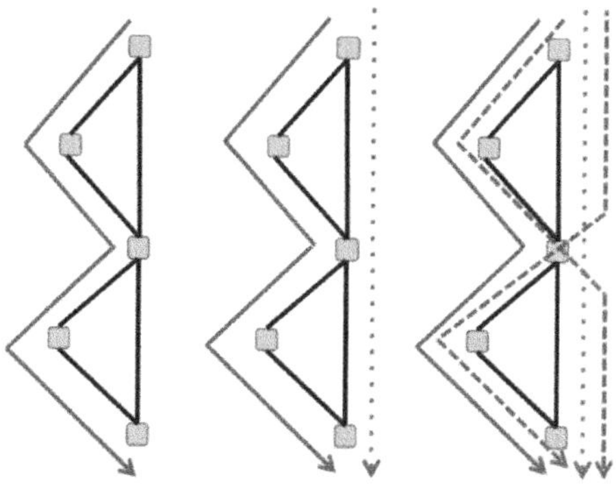

a) b) c)

Figure 4.16. *Three types of coverage*[9]

Coverage of the instructions consists of executing each instruction in the program at least once. Figure 4.16(a) shows that with a single test running through the two "*then*" branches in the program, this coverage is not sufficient because it is designed to test the explicit paths (the absence of an *else* branch introduces an implicit part). Figure 4.16(b) illustrates that the coverage of the branches is designed to test each branch of the execution once. It allows us to see certain behaviors, such as the fact that no instruction is executed (because of the two *else* branches). On the other hand, it does not enable us to verify the program's behavior if the first *then* is not executed and the second is. The risk is that a variable may be updated in the first *then* statement and consumed in the second. Figure 4.16(c) presents the coverage of the paths, and in this

9 The program given as an example here comprises two *if* structures without an *else*.

example this coverage enables us to test all the behaviors and thus detect any errors linked to the various combinations of branches.

An important point is that, for software classed as SSIL1-SSIL2, coverage of the instructions is not sufficient, it is necessary to envisage coverage of the branches.

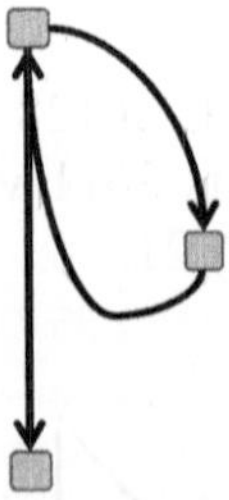

Figure 4.17. *Example of a loop*

Figure 4.17 shows an example of a program comprising a *while* structure (same thing for *repeat* or *for* loops). Coverage of the paths is impracticable on this type of program because the *while* structure must be executed 0 times, 1 times … *n* times. If the *while* structure is composed of a complex body, the number of paths is higher. For this reason, CENELEC 50128 indicates that coverage of the paths is impracticable. Also note that it is possible to focus on coverage of the *i*-paths, where *i* defines the maximum number of loops; when $i = 1$, we obtain a degree of coverage which is achievable in terms of cost.

In terms of coverage of the data flows, the aim is to verify that all the data produced are consumed, and that all the data consumed have been produced. This coverage is not used in industry at present.

Coverage of the instructions, the branches and the paths involves checking the functional progression in the software application. The purpose of data flow coverage is to verify data management. These different types of coverage are unable to guarantee that all the conditions are correctly encoded. Indeed, by going through each branch or path, we can verify that it is attainable, but in order to verify that the conditions (expression evaluated at the level of the IF, WHILE and/or the REPEAT statements), it is necessary to test the conditions themselves.

A condition C is composed of simple conditions – whose value is either true or false. More specifically, the condition C is composed of simple conditions connected by logical operators (and, or, not, etc.).

Consider a condition $C = CS_1$ op1 CS_2 op2 ... CS_n. To fully test the condition C, it is necessary to perform 2^n tests. To perform the tests of the compound conditions (Table A.21 – row 3), it is necessary to test each simple condition, which gives us $2 \times n$ tests. In aeronautics, Modified Condition/Decision Condition (MC/DC) is used, necessitating n+1 tests, as shown by Figure 4.18.

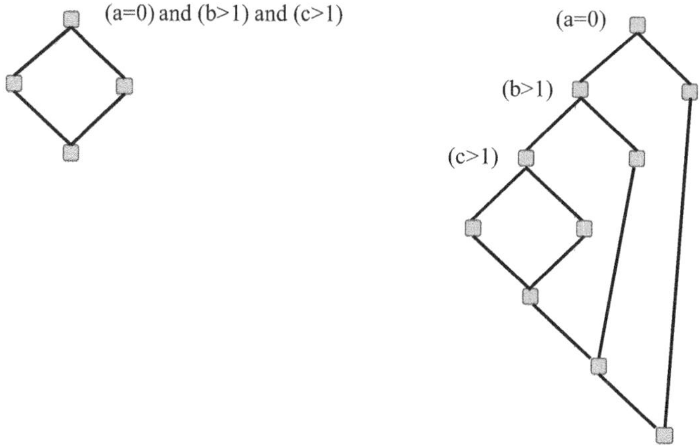

Figure 4.18. *Aliased program*

CENELEC 50128 recommends the putting in place of the compound condition. MC/DC is not the compound condition, but it is based on lazy evaluation as enacted by compilers. As an evaluative measure, this coverage is satisfactory for SSIL3-SSIL4.

As is indicated by item 2 in the requirements in Table A.21, for SSIL3-SSIL4, it is recommended to implement the following combinations when testing the components:

– coverage of the branches + coverage of the compound conditions: all the branches are executed and all the conditions are verified. This combination is interesting, but it may lead to the construction of condition-oriented tests which do not represent the component's true behavior;

– coverage of the branches + data flows: merely covering the branches and the data flows is not capable of demonstrating that all functional paths have been tested;

– coverage of the paths: as previously indicated, this coverage is impracticable when we have repetitions.

In light of the above analysis, we recommend constructing the tests on the basis of the requirements associated with a component and/or on analysis of the algorithms associated therewith. If the algorithms are an input for the specification of the tests, it will be necessary to search for coverage of the conditions but on the basis of scenarios describing a realistic functional framework.

The concept of the level of coverage implies the possibility of possessing information about the execution path associated with each test. The coverage information can be obtained by way of two mechanisms:

– an instrumentation of the code: in this scenario, there is a modification of the source code. This modification may impact on the application's behavior (at least from the temporal point of view). This type of instrumentation is carried out when working on the "host" device;

– a physical instrumentation at thee processor level (a probe), at the bus level, etc. This type of instrumentation is carried out when working on the "target" device.

4.7.2.5. *Summary*

As previously indicated, the activity of verification must be planned, and it must be justified that the approach proposed enables us to detect the various anomalies which could be introduced into the final software. The approach must be verified as being consistent with the company's prescribed processes and with the CENELEC 50128:2011 objectives for the SSIL objective.

The proposed approach must be a combination of dynamic and static analysis techniques. The code must be manipulated and it must be executed. One of the essential points is to render the activity of verification effective. In order to do so, it is necessary to have processes, skilled people and clear objectives. The objectives of verification need to be formalized in control lists.

To conclude this section, we must recap an additional requirement: we must have tangible proof, and therefore the activities of verification must be documented and the data recorded in order to facilitate auditing and independent assessment.

4.7.3. *Validation*

Validation in the context of the lifecycle of a system includes all activities which offer assurance and which build confidence in the system and in its ability to satisfactorily perform the intended functions, and to achieve the set goals and objectives.

As Figure 4.19 shows, validation is an activity which involves demonstrating that the software application actually provides the service for which it was developed. With the intended service being defined by the requirements, it is necessary to show that the software application, running on the target hardware platform, conforms to these requirements, by way of a battery of tests. If we consider the activity of testing as an activity of verification, then validation is the confirmation, by examination and provision of tangible proof, that the software conforms to the specifications given in the software safety prescriptions.

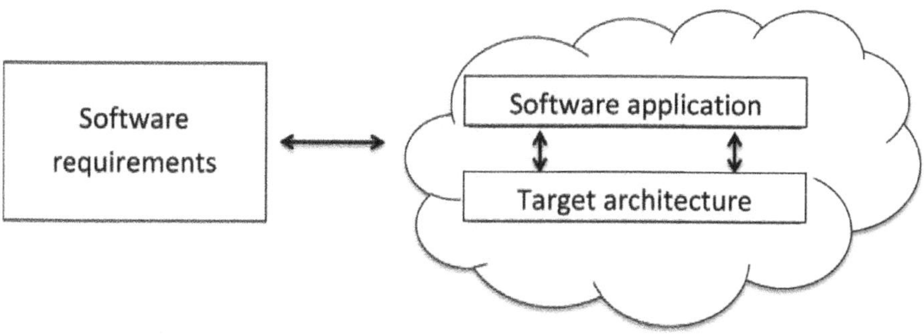

Figure 4.19. *Validation*

CENELEC 50128:2011 introduces the concept of tests performed on the whole software package instead of the term "validation tests". As Figure 4.12 illustrates, all the phases are subject to an activity of verification (a combination of reviews, static analysis, testing, etc.).

Thus, the validation of a software application consists of combining all of the verifications and certifying the fact that the software application does or does not respect the requirements set out in the software specification. As we shall explain in section 8.3.7, a specific report known as the software validation report (SVR) needs to be produced.

DEFINITION 4.3 (Validation).– Confirmation by tangible proof that the requirements for a specific usage or intended application have been met.

According to Definitions 4.2 and 4.3, validation can be seen as external verification of the product.

In summary, validation consists of demonstrating that we have created the product correctly.

4.8. Independent assessment

In both its versions (2001 and 2011), CENELEC 50128 introduces independent assessment. An independent assessment is performed by a person who is not directly involved in the project. He/she analyzes all of the elements produced to realize the software and confirms whether or not the result conforms to requirements and whether or not the SSIL objective has been achieved.

In Chapter 11, we shall give a more formal presentation of the activity of independent assessment and the additional concepts which go along with it, such as that of certification.

4.9. Tool qualification

CENELEC 50128:2001 introduced a requirement for the compilers used to be purpose-certified, but did not give any clear indication of what precisely was expected in this regard. The 2011 version of CENELEC 50128 formally introduces the need to obtain qualification for the tools employed for a project (see section 6.7 of the standard). As with IEC 61508 [IEC 08] and ISO 26262 [ISO 11], three classes of tools are introduced: T1, T2 and T3.

The T1 category is reserved for tools which have no impact on the verification and on the final executable file. T2 is devoted to tools where a fault could lead to error in the results of the verification or validation. T2 contains the tools used for verification of the coding rules, quantifying the metrics, static analysis of the code, management and execution of tests, etc. The category T3 is given over to tools which, if faulty, could have an impact on the final executable software. This class includes compilers, code generators, etc.

Section 6.7 of CENELEC 50128:2011, for each category, defines a set of recommendations which can be used to identify the content of the tool qualification report.

Chapter 9 is devoted to the description of the activities to be carried out to obtain qualification of the tools employed.

4.10. Conclusion

In this chapter, we have presented section 6 of CENELEC 50128:2011, which introduces the concept of software assurance. Software assurance is a concept which introduces the elements necessary for the realization of a software program.

These elements are the management of quality based on ISO 9001:2008, strong and effective verification and validation deployed at each phase, a complete command of the tools and an independent assessment whereby an outside observer verifies that the process is well defined and has been properly applied.

Software assurance has been established to confirm that the realization of a safety-related program involves not the introduction of safety-related techniques (protection, fault detection, redundancy, etc. – see [BOU 13b]) into the software, but rather that it is acquired by way of a firm understanding of quality, V&V, the tools and an independent assessment, with the main idea being that it is crucial to minimize the number of latent errors by introducing as few as possible and detecting as many faults as possible.

4.11. Appendix A: list of quality documents to be produced

In this chapter, we have described the principles of quality management. We now take advantage of this section to recap on the list of quality plans which must be produced as part of a CENELEC 50128:2011-compatible project.

Title	Acronym
Software Quality Assurance Plan	SQAP
Software QA Verification Report	QVR
Software Configuration Management Plan	SCMP
Software Verification Plan	SVeP
Software Validation Plan	SVaP
Software Delivery and Deployment Plan	SDDP
Software Maintenance Plan	SMP
Software Assessment Plan	SAP

Table 4.15. *List of quality plans to be produced*

4.12. Appendix B: structure of a software quality assurance plan

The SQAP for a railway project must contain the elements which are described later on in the section. The SQAP is subdivided into a variety of documents, as shown by Figure 4.5.

The subjects to be dealt with are:

– section 1: identification of the mandatory standards and regulations. The aim of this section is to identify all the standards and regulations which need to be respected during the development of the software. It is necessary to clearly identify the titles, references and versions of these documents. This section includes ISO 9001:2008, CENELEC 50128:20xx, IEC 62279:20xx, etc.;

– section 2: organization of the project (software part), demonstration of independence and justification of the skills (this may be based on a process local to the process and on management of the company's HR);

– section 3: a presentation of the boundary of the program (or programs) being created;

– section 4: a presentation of quality management and control (metrics, control points, audits, etc.);

– section 5: a presentation of the software development cycle (V-model, etc.) and of each step involved. For each such step, there must be a subsection, describing the input elements, the output elements, the activities to be performed, the human resources, the technical resources (tools, testing environments, laboratories, etc.) and the acceptance criteria and end-of-phase management criteria;

– section 6: a presentation of configuration management (tools, procedure, version identification, identification of the elements needing to be managed, etc.). It will be necessary to explicitly define how the software version specification is produced;

– section 7: a presentation of the process of anomaly management, and of demands for modification and corrections;

– section 8: a presentation of the process of tool management (identification, configuration management, etc.), and especially the management of tool certification;

– section 9: we must have a list of the documents which need to be produced during the software development process;

– section 10: it is necessary to demonstrate conformance to the standards identified in section 1 – specifically to CENELEC 50128 (or IEC 62279). Conformance to CENELEC 50128 should not be based solely on the tables given in Appendix A, but rather on the whole of the standard. Indeed, the body of the standard is normative, and there are certain subjects which are not covered by the Tables A.x. For example, in the 2001 version, the coverage of the tests is not discussed by a Table A.x, but is dealt with by a sentence within the main body of the document. For the 2011 version, it is the certification of the tools which is not dealt with by a Table A.x.

5

Requirements Management

5.1. Introduction

Requirements engineering is a need which appears in all professional standards (in the fields of aeronautics, automobiles, railway, the nuclear sector, electrical equipment, etc.). A difficulty is introduced by the fact that none of these industrial standards actually define what is meant by a "requirement". The industrial standards introduce the concept of traceability (a link between different elements) and that of a level (in aeronautics we have the concepts of High-Level Requirement and Low-Level Requirement (HLR and LLR)).

This chapter presents the activities of requirement engineering and their implementation. The activities identified cover the user needs analysis and the realization.

Requirements management is the discipline which consists of establishing and documenting requirements. The various activities associated therewith are elicitation, specification, analysis, verification and validation, and management.

As a general rule, a project will begin with a phase of requirements acquisition, the purpose of which is to construct a requirements specification. On the basis of the requirements specification, the second phase is to realize those requirements.

In reality, certain projects begin with the phase of analysis; others will not. Indeed, a client lodging an order who wishes to create a new piece of equipment must begin with an analysis phase, in order to construct the technical specifications list as accurately as possible.

For other projects, it is considered that the requirements specification exists and even that it is provided in the form of the technical system needs (TSN). It should be noted that this second category of projects corresponds to the fact that a client lodging an order, or a high-level industrialist, will use several companies to create the final system, and in this case, the requirements specifications serve to ensure the consistency of the whole system.

5.2. Requirements acquisition phase

5.2.1. *Introduction*

The first phase (see Figure 5.1), which can be called the requirements acquisition process, is a four-step process (see Figure 5.2): requirements elicitation, requirements analysis (negotiation), requirements document, and verification.

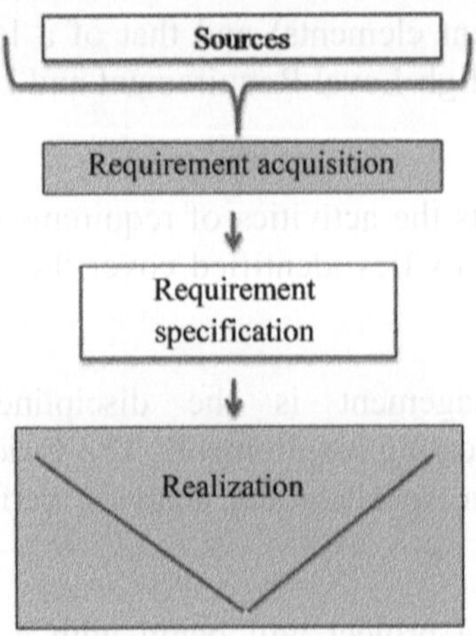

Figure 5.1. *Two-phase process*

The purpose of the first step, called elicitation, is to identify the problem (identification of the stakeholders, sources, explicit and implicit requirements). The second step involves analyzing the problem, discussing and negotiating. This second step is very important, because it is necessary to have the agreement/assent of the user(s) as regards the requirements; here the notion of negotiation is very important in arriving at a consensus.

The next step is to produce the requirements specification, which may be in test format, and/or be supplemented by models. The final step is the verification of requirements, on the basis of the requirements specification: it is necessary to verify that the specification is coherent and complete.

The realization phase is based on a cycle which must integrate all the different phases of the system's life.

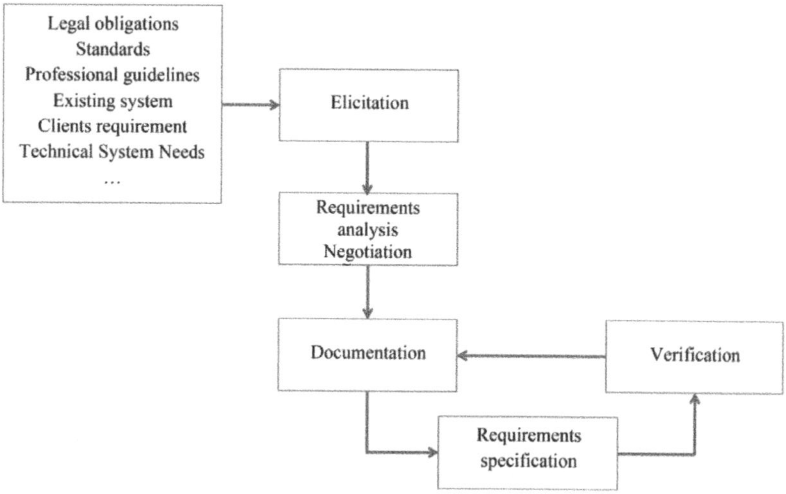

Figure 5.2. *Acquisition process*

In Chapter 2 of [RAM 09] and Chapter 3 of [RAM 11], examples of requirements management in the automotive and railway domains are presented.

5.2.2. *Requirements elicitation*

5.2.2.1. *Introduction*

Table 5.1 is an extract from [STA 94]. It shows that over 30% of causes of failure in the realization of systems stem from incompleteness of the requirements, a lack in the description of the requirements or unrealistic requirements.

Description	%
Incomplete requirements	13.1%
Requirement not representative of the user's needs	12.4%
Poor resource management	10.6%
Unrealistic requirements	9.9%
Poor support from the management	9.3%
Evolution of requirements/specification	8.7%
Poor planning management	8.1%
Not a requirement	7.5%

Table 5.1. *Distribution of the causes of failure*

One of the difficulties of requirements management lies in defining the concept of a requirement. There are many published works which attempt to identify just what a requirement is, and how to cater for it. [HUL 05] offers one of the most complete overviews.

We shall adopt definition 5.1, which is drawn from work carried out by industrialists in AFIS[1].

DEFINITION 5.1 (Requirement).– *A requirement is a statement which expresses a need and/or constraints (technical, cost-related, time-related, etc.). The language used for this statement may be natural language, mathematical language, etc.*

Attributes	Description
ID	Unique Identifier
TEXT	Text of the requirement
SOURCE	Element from which this requirement is derived

Table 5.2. *A requirement*

1 AFIS is the *Association française d'ingénierie système* (French Systems Engineering Association). One of its working groups is specifically devoted to requirements management. For further information, see: www.afis.fr. AFIS is a subsidiary of INCOSE – see: www.incose.com.

For clear identification, a requirement is a labeled element (the label provides a unique identification) that characterizes an element of the system under construction. For each requirement, it is important to know the source, so the attribute "source" is included. Table 5.2 characterizes what is meant by a minimum requirement, an identifier, a text and a link to the source.

Requirements elicitation consists of identifying, clarifying and justifying the requirements which need to be integrated. One initial difficulty lies in the fact that there are numerous sources of requirements (see Figure 5.3): they include the client's specifications, the professional guidelines, the existing system, similar system, interface systems, applicable standards and laws, user needs, etc. The second difficulty lies in the fact that the initial expression of the needs is often incomplete and extremely vague.

With documentary sources, the person in charge of elicitation of the requirements (hereinafter called the "analyst") will use all sorts of documents, such as the existing documentation about the product, bug reports written about previous applications, documents from monitoring (or "vigil") of the sector (standards, laws, etc.) and technological environment (new technologies, etc.) or analysis of previous change requests.

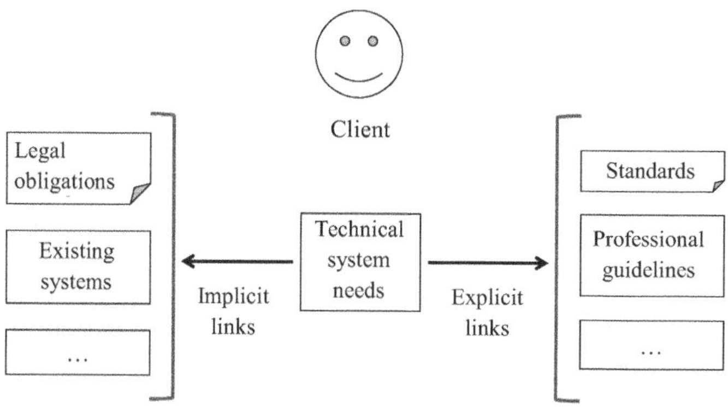

Figure 5.3. *Sources of need*

Yet completely focusing on the end user (who will need to be clearly identified) will enable the analyst to gather the user's expectations and remarks regarding the functional aspects, on the one hand, but most importantly the non-functional aspects. Particularly long response times for an application, problems of accessibility or ergonomics will always be subject to comments, which then need to be archived for later use.

Ultimately, it is fairly difficult to find the optimal combination of all of the resources at the analyst's disposal. The task of requirements elicitation entails:

– identifying the stakeholders;

– identifying all the sources of the requirements as far as possible;

– adapting the analysis strategy to the particular problem at hand (systematic analysis of the documentation, immersion with the client, immersion with the end user, brainstorming sessions, etc.).

These three activities will be analyzed in this chapter. There are a number of different types of requirements, as illustrated by Figure 5.4.

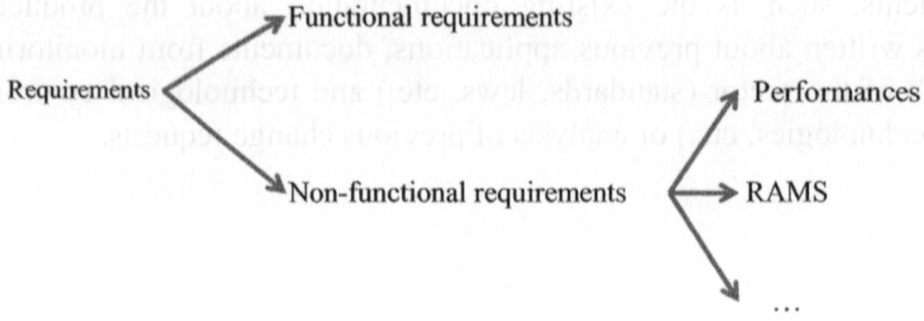

Figure 5.4. *Acquisition process*

The requirements can be classified into two categories: functional requirements and non-functional requirements. Functional requirements pertain to the system's behavior, and can therefore be tested, whereas non-functional requirements characterize properties such as safety (causing of harm to people, goods and/or the environment), security (respect of the labor codes, intrusion, data manipulation, etc.), availability, reliability, performances, maintainability, etc.

5.2.2.2. *Identification of stakeholders*

As a general rule, projects tend to be to update an existing product, so there is a feeling of familiarity with all the stakeholders involved in the project, and in the rarer case of innovative projects, there is a certain feeling of freedom. Therefore, the step of identification of the stakeholders is often overlooked. This oversight has an impact on the construction of the requirements framework, which results in an incomplete fulfillment of the needs.

For so-called certifiable systems, this may cause a lack of knowledge or incomplete knowledge of the applicable laws and standards. Hence, forgetting the stakeholders is often a cause of failure or significant delay in projects, because the products delivered do not correspond to the expectations held by the stakeholders (operators, network and/or infrastructure managers, maintenance staff, users, authorities, certification bodies, etc.).

A stakeholder is, by definition, a person or entity who has an interest in the project. One way to gain a complete picture of who the stakeholders are is to attempt to answer the following questions:

– who is financing the project? Who is in control of the budget?

– does this project have sponsors? If so, who are they?

– who are the users of the future system? By "user", evidently, we mean the end user, but not solely the end user. We must think about the system's time in operation (the operators, administrators, etc.), its maintenance and its eventual decommissioning. Figure 5.6 shows a number of users (operators, motorists, train drivers and maintenance staff);

– who are the people that will design and test the system?

– who will be installing the system and training the end users?

– are there any authorities in charge of authorization of the system? There may be national, regional (e.g. European) and international bodies, each of which will have applicable legislation and representative organizations;

– are there any pre-existing interface systems?

– is there a need for certification?

– etc.

It is possible to analyze the needs pertaining to the stakeholders by creating a model of the environment in its true context and using it to identify the flows between the parties. In order to do so, we can create a context diagram or an environment diagram in the form of a "horned beast" or "octopus" diagram, as used in the APTE[2] method [BRE 00].

In a project consisting of linking the various pieces of equipment in the railway system by a global network, we can construct an octopus diagram identifying the new system and the actors who will interact with it. Based on that diagram, it is possible to identify the crucial functions (without these functions, the system cannot operate) and the main functions (required services). Figure 5.5 illustrates the result of this study.

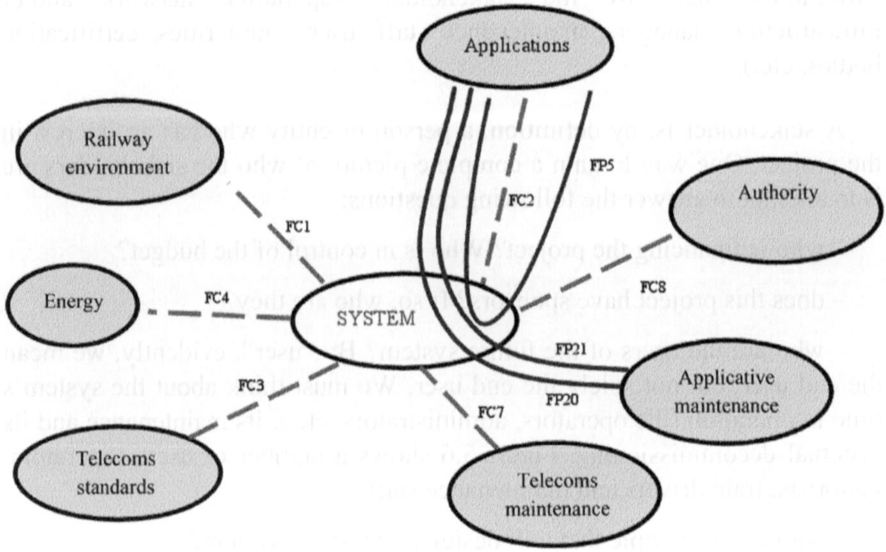

Figure 5.5. *Example of an "octopus" diagram*

Figure 5.6 is an example of a model (class diagram) that can be employed to identify the users and the interactions between those users. This model uses UML notation [OMG 11, ROQ 06, ROQ 07] and demonstrates that the Dynamic Radio-Based Control System (DRBCS) being designed is a decentralized system which interacts with several types of actors (the operators, train drivers, maintenance personnel and road users).

---

2 The APTE method is a functional-analysis and value-analysis method. This versatile method can be applied equally for products, manufacturing procedures, equipment or organizations. For further information, see: http://cabinet-apte.fr.

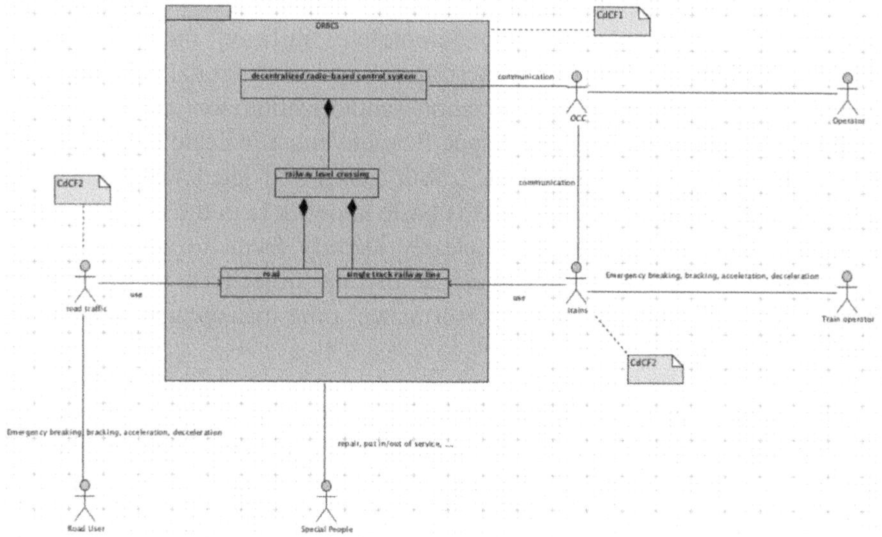

Figure 5.6. *An automated level crossing and its users*

It is important to identify each of the stakeholders and clearly indicate what their interest is in the project. In general, a table with columns (for example, Table 5.3) serves this need to identify the stakeholders.

Name	Organization	Roles	Contact details	Availability	Field	Level of expertise	Objective, Interest in the project

Table 5.3. *Example of identification of stakeholders*

5.2.2.3. *Identification of sources*

Generally, the project will be a response to a demand (usually in the form of an updated set of functional specifications) which is expressed by one of the stakeholders: the client. Hence, the functional specifications are the first source.

The functional specifications may contain domain-specific requirements (e.g. specific to the aeronautics, automotive, railway, nuclear, service, telecoms domain, etc.) and/or make reference to professional documents. If the product is subject to authorization (authorization for commissioning, certification, etc.), the functional specifications must indicate the applicable standards (IEC[3], DO[4], CENELEC[5], ISO[6], etc.) and legal texts (laws and decrees) to be respected, unless this is basic knowledge in the domain, but as a general rule, it is preferable to clearly identify them. In addition to the standards and legal texts, we must define the objectives to be fulfilled (objectives pertaining to safety, performance, time management, maximum workload, certification, etc.).

On the same principle, if the system is intended to be a replacement for an existing system, the documentation for the existing system and the feedback on that system also become sources which need to be taken into account.

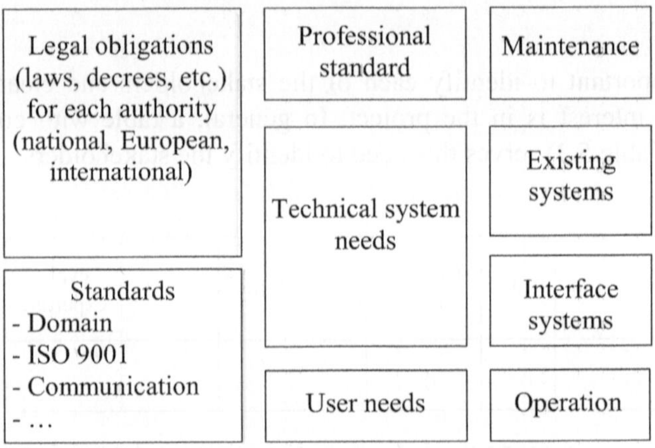

Figure 5.7. *Identification of sources*

3 See: www.iec.ch/.

4 Standard published by the RTCA inc. See: www.rtca.org/.

5 See: www.cenelec.eu/.

6 See: www.iso.org/iso/home.htm.

It should be noted that in the railway domain, the "GAME[7]" (*Globalement Au Moins Equivalent* is the European term for Globally At Least Equivalent, see [LET 00]) principle is used in the design of a new system. With this principle, it is possible to draw on past experience (reuse of justifications, recycling of the initial principles, etc.) in the design of a comparable system.

More generally, the functional specifications are incomplete, and on the basis of the list of stakeholders, it is necessary to identify the associated sources. Figure 5.7 gives an illustration of what may constitute the sources for a project.

5.2.3. *Process of analysis and documentation*

5.2.3.1. *Subdivision of the process*

On the basis of the identification of the stakeholders and knowledge of the sources, it is then possible to implement an analysis process (see Figure 5.8) which can be divided into two steps: the analysis of the problem (or requirements analysis) and the production of the description of the product.

This two-phase process is important, but there is a tendency, with many projects, to go directly to the production phase. In this case, there is a danger of not fully grasping all of the needs, and therefore creating the wrong system or an inadequate system.

The analysis phase involves identifying the requirements. In order to do so, we can analyze all of the sources and select the known requirements, and then on the basis of the list of stakeholders, it is possible to put in place an operation of attribution of the requirements to the stakeholders. This phase will be followed by an analysis phase, the aim of which is to analyze the

---

7 In the railway domain, a positive approach was adopted, stipulating that all new systems or all modifications made to an operating system should offer a global safety level at least equivalent to that of existing systems offering comparable services ("GAME" principle). Such an approach effectively has the intrinsic merit of maintaining the dynamics of progress by minimizing the risk of the "always more" approach. It enables us to take advantage of the existing systems, deemed satisfactory, and thus benefit from the experience acquired. Finally, this approach is essentially determinist, for it is based, as far as possible, on existing reference systems, but for all that, it does not systematically reject the probabilistic approach which is, in any case, necessary in the case of technologically-innovative subsystems.

interactions between the stakeholders and the system, with a view to identifying the additional requirements.

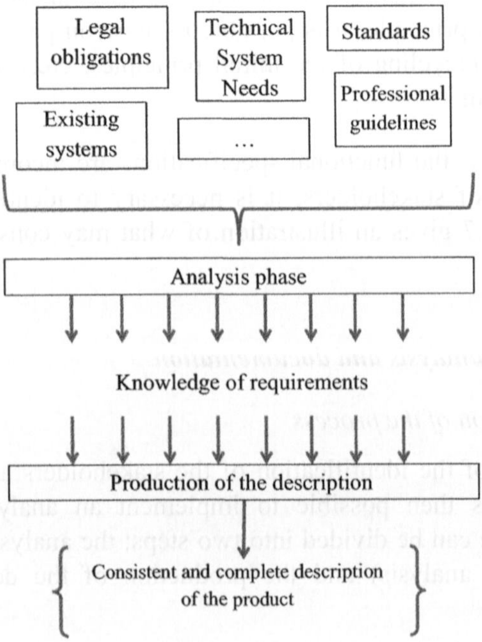

Figure 5.8. *Analysis process*

5.2.3.2. *Requirements analysis phase*

5.2.3.2.1. Objectives

The objective of the requirements analysis phase is to construct all the stakeholder requirements. It should be noted that analysis phase helps to identify the boundary of the product (see Figure 5.9) and to characterize the interfaces with the other products.

Figure 5.9 illustrates the environment of a system, and/or an application, which comprises three inputs (E_i), two outputs (S_j) and three interfaces (I_k) with existing resources (reused systems, electricity supply, etc.).

The stakeholders' requirements form the basis of the acceptance or non-acceptance of the system, the negotiation and agreement on the project, the development of the system and the management of changes made to the requirements. The requirements define the result as expected by the stakeholders. That is why it is necessary to acquire the users' needs as fully as possible.

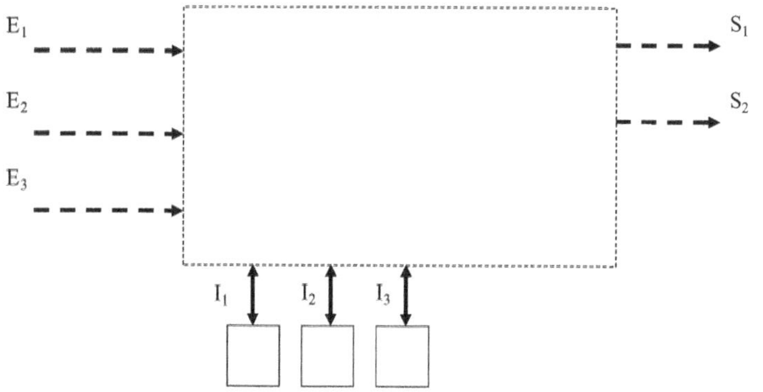

Figure 5.9. *Environment of the software application*

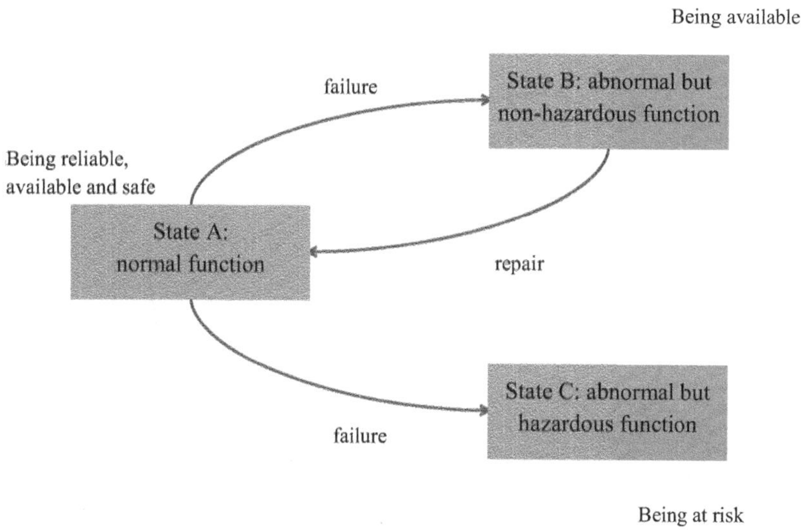

Figure 5.10. *Evolution of the state of a system*

Thus, within the system, we need to identify:

– the interfaces with the environment (see Figure 5.9). These interfaces may be electrical, mechanical, software-based, hydraulic, etc.;

– the states: at rest, in operation, downgraded mode, etc. (for example, see Figure 5.10). The concept of the state introduces a division between the states of normal operation, states of non-operation and hazardous states;

– the concept of correct behavior, downgraded behavior and hazardous behavior;

– the concept of functional and non-functional needs.

As regards the states of the system, Figure 5.10 identifies the proper and improper states, but it is necessary to go further and introduce all of the states that may be encountered for the system, which characterize specific behaviors (init, safe state, maintenance, degraded, etc.), as illustrated by Figure 5.11.

During the elicitation phase, it is necessary to reflect on the non-functional requirements and implement analyses relating to the operational safety, the aim of which is to define the safety requirements but also the availability, reliability and maintainability requirements.

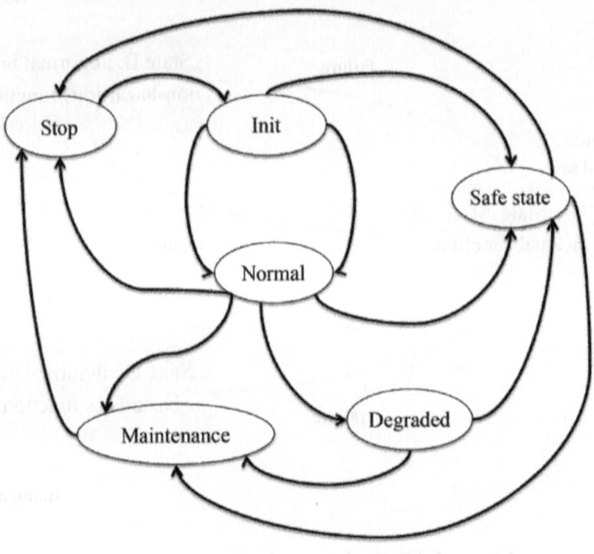

Figure 5.11. *Different states of the system*

Figure 5.12 illustrates a process that integrates the elicitation of the non-functional requirements pertaining to RAMS.

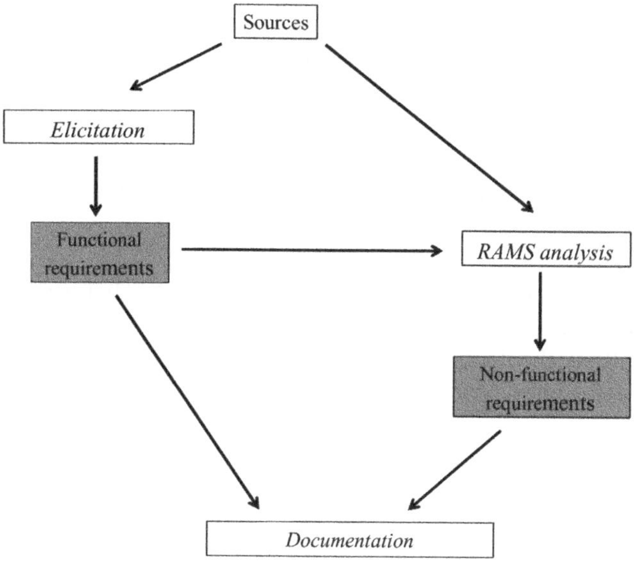

Figure 5.12. *Elicitation process with RAMS analyses*

5.2.3.2.2. Elicitation techniques

The purpose of elicitation techniques is to aid the discovery of the conscious, unconscious and subconscious requirements of the stakeholders. These techniques are chosen as a function of the risk factors, the human and organizational constraints, the professional domain and the level of detail expected for the requirements. The elicitation techniques may also be chosen as a function of the requirements document being prepared.

On the basis of the list of stakeholders and the sources, it is then necessary to establish a process of acquisition of the requirements, and with that goal in mind, there are a variety of elicitation approaches [ZOW 05]:

– inquiry techniques: surveys, questionnaires, etc.;

– interview techniques;

– creativity techniques: brainstorming, storyboarding;

– animation techniques: brown paper, roleplaying, use cases;

– observation techniques: terrain, learning;

– prototyping and simulation techniques.

In Appendix A of [MEI 02], the author presents the methodological elements pertaining to requirements management. Requirements management is one of the tools which system engineering offers, as is demonstrated by the EIA-632 standard [EIA 98, EIA 03].

The best result is obtained when the analyst jointly implements several of these techniques.

In this book, it is not possible to be exhaustive. Therefore, in the next sections, we are going to present only two of these techniques, but it must be noted that all of the techniques identified above need to be implemented, depending on the particular project at hand.

5.2.3.2.3. Interview techniques

An interview comprises several steps:

– question phase: the questions are posed and the responses noted;

– pause: during pauses, most people will find something to say and/or explain, which offers the opportunity to discover the additional requirements;

– summary and/or reformulation phase: this phase is important because it enables us to verify that the responses have been fully understood.

The implementation of interviews involves a preparation phase, where a set of questions will be prepared. This set of questions should enable the interviewers to identify the needs of the different stakeholders. This set of questions contains:

– open-ended questions: these require responses other than “yes” or “no”;

– closed-ended questions: such questions can be answered with “yes” or “no”. Closed questions are used to obtain a definitive response (after reformulation, for example).

It should be noted that during the interview, it is possible to add questions depending on the responses and after the reformulation of the responses.

It is necessary to interview all the stakeholders who have been identified, and make them aware of the fact that their requirements form a whole (a system). It is crucial to take the stakeholders seriously and not to pass judgment as to the requirements they express. During the interviews, it is necessary to treat all of the elements as requirements, to document the importance of the requirements for each stakeholder, document the results of the interviews and formalize the stakeholders' acceptance of the result (notes, documents, etc.) of the interviews.

The use of interviews involves stimulating and inciting the stakeholders to respond. The chairperson of the interview panel therefore needs to have excellent abilities to manage a discussion and sustain it, whilst being capable of managing the pauses.

5.2.3.2.4. Prototyping and simulation techniques

A maquette is an initial, highly-simplified model of the problem. The maquette is designed to model certain aspects of the problem to be solved in a more or less precise way. When the maquette handles genuine elements of the system (such as the data files, etc.), we speak of a prototype.

The maquette and/or prototype is a way of dealing with the vision of the different players and the behaviors expected (see Figure 5.13).

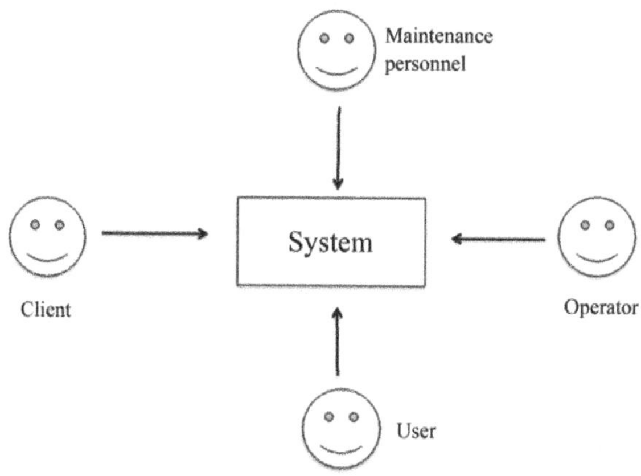

Figure 5.13. *Prototype and expression of needs*

A prototype may be static, in which case we seek to model the interactions between the different elements, including the actors, but it may also be dynamic, and we model simplified behaviors. In that case, it will be possible to run scenarios.

The realization of a prototype is a good way of facilitating the understanding of the needs, but this involves costs, and often it is tempting to consider that the prototype is the start of a solution.

It should be noted that more and more prototypes are used to validate a concept (Website, etc.). The difficulty with a prototype is the need to not lose sight of the fact that it *is* a prototype (which may be extremely advanced) and that it is not the design for the final product.

5.2.3.3. *Description production phase*

The second phase consists of identifying the requirements. The process of analysis and transformation is designed to clarify the text of the stakeholders' requirements and identify the product requirements. The conflicts (two contradictory requirements or requirements with different objectives), any incompleteness, unspoken requirements and others need to be demonstrated.

It is at this step that it is appropriate to retire the descriptive aspects and focus on what is essential. Indeed, it is necessary to focus on the need and not on pseudo-solutions. At the conclusion of this phase, we obtain a set of requirements which constitutes the description of the needs.

Finally, we seek to produce a complete and consistent description of the requirements. As regards the requirements specification, we describe the activity in section 5.3.

5.2.4. *Verification and validation of the requirements*

5.2.4.1. *Introduction*

In Figure 5.2, we have identified the necessity of verifying the requirements following the phase of production of the requirements specification document. However, we also need to speak of the concept of validation: indeed, the requirements identified with the client will be the source for the client tests, which are usually called "acceptance tests".

The realization of a system (subsystem/equipment/application) must take account of the design, but it must also take account of the activities which prove that the system has achieved a certain level of quality. The achievement of a level of quality involves demonstrating that no faults have been introduced during the design phase and that the product corresponds to the requirements which have been identified.

In the context of the requirements and the phase of verification of the requirements specification, it is necessary, first of all, to demonstrate by means of a verification phase, that the requirements specification does not contain errors, and by way of validation of the requirements that the system produced does indeed correspond to the client's needs (see Figure 5.14).

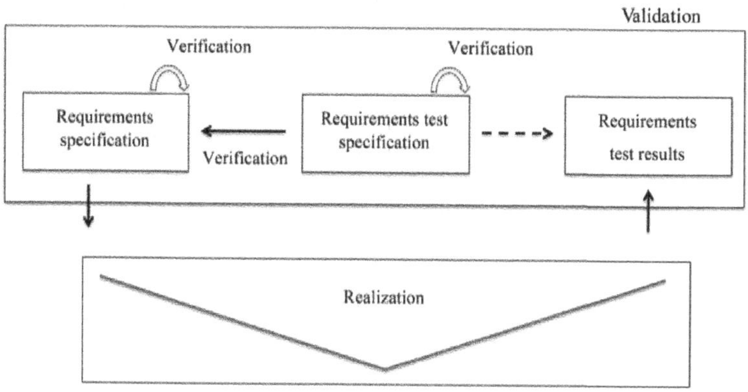

Figure 5.14. *Verification and validation*

5.2.4.2. *Verification*

The verification of the requirements can be performed by way of several activities:

– more or less formal design review (readthrough, use of checklists, etc.);

– realization of a model and/or a prototype;

– preparation of test log (see Figure 5.15).

It should be remembered that here we are speaking of a verification that the stakeholders' requirements have been properly taken into account. Thus, it is necessary for this task of verification to be performed jointly by the client and the supplier.

The validation of the requirements involves the creation of a battery of tests to ensure the system is accepted by the client. This is one of the most important stages of verification in the whole process.

As Figure 5.15 illustrates, on the basis of the requirements, we identify test cases[8] (TC_x). These test cases describe a particular situation to be achieved, with this situation being linked to an equivalence class. Based on the test cases, it is possible to prepare the test scenarios; each test scenario describes a situation. Thus, a test case may play a part in several test scenarios.

As the only element of input at this level is the requirements specification, the tests we are describing here are what are known as "black-box" tests (as we have no knowledge of the realization).

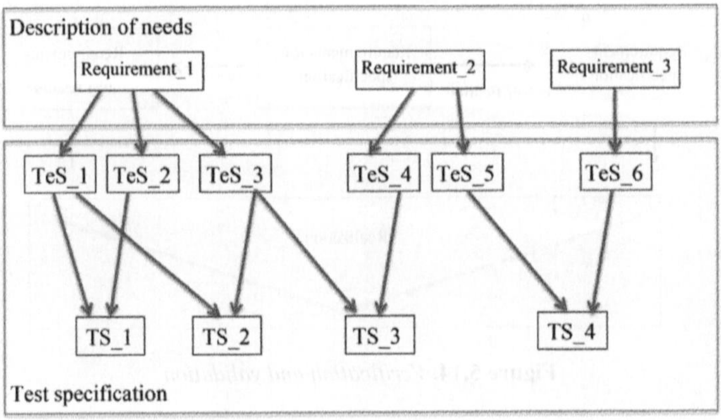

Figure 5.15. *Link between the tests and the requirements*[9]

The client, or the entity working on behalf of the client, usually must perform the acceptance tests, but in view of the contract, the client's history, the supplier's history, etc., the client may decide to construct their own battery of acceptance tests on the basis of the validation tests performed on the installation, conducted by the supplier.

5.2.4.3. *Validation*

In order to conduct the acceptance tests, we need to have access to the system in its true environment. This is an important activity, because the aim is to verify that the finalized, installed product functions as the client

8 For more information on the testing, see [MYE 10].

9 TeS for Tests Specification and TS for Test scenario.

intended. This phase is also known as the receipt phase, and in the CENELEC 50126 standard [CEN 99], it is called the system acceptance phase.

5.3. Requirements specification

5.3.1. *Requirements characterization*

The requirements specification is a document detailing all of the requirements to which the product must conform. Definition 5.1 defined the concept of a requirement, and Table 5.2 defined the fundamental attributes which characterize a requirement.

5.3.1.1. *Identification*

A requirement must be uniquely identifiable, so generally an identifier is attached to each requirement. This identifier must be unique. So as not to cause confusion with other items which may be numbered (such as items on a list, etc.), the identifier is constructed using a label (e.g. REQ) and a unique number (xxxx). Thus, we have an identifier in the format REQ_xxxx.

[REQ_0001]

The software must be able to be updated.

[PROJECT] AAAA

[DOCUMENT] DDD

A requirement is associated with a project, and in particular with a document. Hence, it is possible to have an identifier in the format REQ_AAAA_DDDD_xxxx, where AAAA represents the project number, DDDD the document reference and xxxx the unique number assigned to that requirement.

The project reference and document reference can therefore be integrated into the identifier or be seen as an attribute.

[REQ_AAAA_DDD_0001]

The software must be able to be updated.

If we want to make the identifier unique, it is important not to reassign the requirement numbers if a requirement is removed. For that reason, it is wise to avoid using an automated identification mechanism (such as the auto-numbering style in Microsoft® Word).

As regards the identification of a requirement, it may be difficult to detect the end of a requirement. Therefore, it is useful to put an identification system in place in the form of tags – see the following example.

[REQ_0002]

A software cycle must be executable in the space of 100 ms.

[END_REQ]

5.3.1.2. *A few important characteristics*

Before defining any requirements, it is necessary to put a set of criteria in place. These criteria should be used to qualify the requirements. An analysis of the existing literature (academic and/or normative) enables us to identify criteria which pertain to a requirement, and others which pertain to a set of requirements.

For each requirement, the following criteria are most often encountered:

– atomic: the requirement is a clearly-identifiable and indivisible element; it expresses one point and one point only;

– concise: when it is described in natural language, a requirement must be written in the form of a single sentence, which should be no longer than a few lines;

– clear: the requirement can be fully understood on the basis of a single reading; the sentence-structure is simple and does not use literary flourishes;

– precise: all the elements used in the requirement are identifiable and fully characterized (no pending questions such as: what units of measurement are used?);

– abstract: a requirement is at the appropriate level of abstraction: it should not impose a solution (technical or functional), but merely describe the need;

– unambiguous: reading of the requirement should facilitate an understanding of the need with only one possible interpretation. Thus, it is important not to use turns of phrase or words which facilitate multiple interpretations or complexify the understanding of the requirement;

– up to date: the requirement must reflect the current state of the system / of the knowledge about the system;

– complete: it is key that all of the concepts used in the requirement be defined and that no information be lacking;

– verifiable: there must be a means of verification of the requirement which is feasible (within reason);

– consistent: it is necessary to verify that the requirement forms a consistent whole (i.e. the terms used in the requirement are the same throughout and carry the same meaning);

– coherent: the requirement must not be conflictual (i.e. the requirement should not say both one thing and the opposite), and the terminology must be coherent (with the glossary);

– correct: it must correspond to a real need on the part of a stakeholder (external coherence);

– traceable: it is necessary to be able to trace the source, evolution, impact and use of the requirement.

Other criteria may be necessary – e.g. the fact that a requirement should be realizable, meaning that an implementation of that requirement should be possible within reasonable conditions of cost, time and realization.

Ultimately, even though other criteria may be added to this list, those presented above constitute a solid basis upon which to build a methodology.

There are two main rules for the comprehension of the requirements:

– use short sentences and paragraphs;

– formulate one requirement, and one only, per sentence.

5.3.1.3. *Characterization of a set of requirements*

For a set of requirements, the main criteria relate to their coherence, their completeness and non-redundancy.

A set of requirements must be:

– complete: no requirements should be missing and each of the requirements included in the list should be complete. The completeness of the set of requirements is a tricky point, because it is linked to an exhaustive needs analysis. For example, it is easy to forget, in a set of technical specifications, to define the software's behavior in case of undesirable events (such as a hardware breakdown, an error in the data input by the user, etc.). In such situations, it is not for the developer to devise what the program must do at the point of implementation. It is necessary, during the identification of the requirements, to check that the requirements cover:

- all of the objects handled,

- all of the possible states of those objects,

- all of the use conditions,

- all of the use scenarios which have been envisaged,

- all of the applicable standards and professional guidelines, etc.

The best way of evaluating the completeness of a set of requirements is to put a document model in place, which is:

– coherent: there must be no ambiguity or internal inconsistency of the requirements framework; there should be no contradictions between the requirements, and a unique identification of the document. The coherence of a set of requirements relates to the clear definition of the concepts for all of the requirements. Put differently, every word must be used in exactly the same way for each requirement;

– non-redundant: it is necessary that, within the set of requirements, there be no redundancy: the same information and/or the same requirement should not appear multiple times.

5.3.1.4. *Characterization of the process*

From the point of view of the process, all requirements must be:

– identifiable: it is necessary to attach, to each requirement, a unique identifier (see the first attribute in Tables 5.2 and 5.4);

– verifiable: it is necessary to ensure that all the requirements are verifiable. The form of verification may be a readthrough, a model, a specific analysis and/or a test (see the fourth attribute in Table 5.4);

– modifiable: we must be able to manage the evolutions of the requirements throughout the whole of the system's life (realization, manufacture, commissioning, maintenance, decommissioning). In order to do so, a configuration management process needs to be established (see the fifth attribute).

Attributes	Description
ID	Unique identifier
TEXT	Text of the requirement
SOURCE	Element from which this requirement is derived
VERIFICATION	Activity of verification associated with the requirement
VERSION	Version associated with the requirement

Table 5.4. *List of attributes characterizing a requirement*

Table 5.4 introduces a second identification of the attributes describing a requirement. We shall continue to add to this description as the chapter continues.

5.3.1.5. *Additional*

It is thus possible to supplement this list of attributes with other attributes that can be used to qualify the requirement:

– category (functional, RAM[10], safety, performance, etc.);

– status of the requirement: under construction, awaiting validation, validated, implemented, etc.;

– priority (to be defined depending on the project);

– verifiable (yes/no);

– type of verification (readthrough, specific analysis, simulation, etc.);

– testable (yes/no);

– type of tests;

– source (who, when, etc.);

– status (pending, analyzed, rejected, etc.);

10 RAM for Reliability, Availability and Maintainability.

– type of document;

– version;

– effort;

– attribution.

Attributes	Description	Values
ID	Unique identifier	REQ_AAAA_DDDD_xxxx
TEXTE	Text of the requirement	
SOURCE	Element from which this requirement is derived	
CATEGORY		FUNC, RAM, SAFE, PERF
VERIFIABLE		YES / NO
VERIFICATION	Activity of verification associated with the requirement	Walkthrough Design review Calculation Unit tests Integration tests Whole tests …
…		
VERSION	Version associated with the requirement	xx.yy

Table 5.5. *List of attributes characterizing a requirement (cont.)*

The attributes need to be defined at the beginning of a project, and it is necessary to define them very clearly; see Table 5.6.

Name of attribute	Version
Semantics of attribute	Version associated with requirement
List of possible values	x.y
Semantics of values	x is the major index y is the minor index
Unit	N/A

Table 5.6. *Definition of an attribute*

5.3.2. *Characterization of requirements specification*

It has already been mentioned that a set of requirements must be non-redundant, coherent and complete, but the requirements specification document must satisfy additional criteria:

– the document must be structured: that is, it must have a clear structure, which can be easily read by anyone, and which facilitates targeted reading;

– the document must be modular: the requirements which belong together should be grouped in a clear structural framework, and located appropriately in relation to one another;

– the document must be extendible (i.e. the possibility of maintenance must be taken into account): it must have a flexible structure in order to facilitate change;

– the document must be sufficient: there should be no need to wade back through all of the source documents in order to comprehend the product;

– the document must be traceable: relations must be established between the requirements documents and the other engineering documents.

5.3.3. *Expression of requirements*

5.3.3.1. *Natural language*

It is crucially important to express the requirements in natural language. Indeed, natural language remains the most widely used and the simplest of ways to communicate about the requirements and establish mutual understanding. The purpose of expressing a requirement in natural language is to express the need as clearly as possible, without going into the details of the design.

It should be noted that this approach may seem contrary to the common approaches used nowadays, in which the model tends to be central. However, the making of a model cannot replace the expression of the requirements in natural language. Indeed, whilst a picture is worth a thousand words, it is no easier to understand, and it is important to employ those thousand words in order to verify that the picture is correct and complete.

The difficulty of natural language lies in the ability to express the same concept in a number of different ways, and the tendency to introduce literary turns of phrase which render a sentence more complex and may cause ambiguities to arise. Therefore, it is necessary to define rules to govern the description of the requirements in natural language.

The first rule is to use and respect a requirements template, which will need to be defined for each project. This template may be based on the following (non-exhaustive) list of rules:

– a requirement must be in the form "subject + verb + complement";

– consistently use the verb "must + auxiliary", correctly conjugated;

– use the active rather than the passive voice;

– use terms with an unequivocal definition. You can ensure the terms are unequivocally defined by constructing a glossary (see next section);

– avoid the use of adverbs which make the sentence unclear;

– avoid negations. Their use must be limited to the so-called safety requirements (such as "the system must not…").

Below are a few examples of textual requirements:

[REQ_1]

The objective of a firearm is to propel a bullet in a given direction.

[END_REQ]

[REQ_2]

A bullet will be expelled from the chamber if the firearm is not blocked and the powder is ignited.

[END_REQ]

5.3.3.2. *Construction of a glossary*

The difference in interpretation of the terms used by the stakeholders is a common source of conflict and misunderstanding in requirements management. It is particularly important to reach agreement as to shared terminology.

With this in mind, it is possible to define all the relevant terms in a glossary. That glossary would contain domain-specific jargon and technical terms specific to the context, abbreviations and acronyms. If necessary, with a view to making the connection between the stakeholders, the glossary can also identify synonyms and homonyms, but the idea is to define one unique set of vocabulary for the project.

The use of a glossary must conform to strict rules:

– the glossary must be centralized and the version specified: it is a referential standard;

– the responsibilities pertaining to the maintenance of the glossary must be defined;

– the glossary must be maintained and be accessible throughout the duration of the project;

– it is imperative that all the stakeholders make use of the glossary for any communications;

– the glossary must be approved by all of the stakeholders.

It is recommended to establish the glossary as soon as possible, because any misunderstandings can lead to tasks being repeated, but also, in particular, to difficulties in comprehension which may not be discovered until late on, and which may have a major impact in terms both of effort and cost to rectify.

5.3.3.3. *Formalization*

The expression of the requirements in natural language comes up against a problem, which is the declarative nature of the requirements: it may be rather difficult to express a requirement in text form, or at the very least, that textual formulation may be longer and more complicated than a small diagram or a mathematical formula. For this reason, a requirement expressed in textual form can be supported by graphic representations and/or mathematical expressions.

Hence, it is not uncommon to find requirements which are a mixture of a textual description and a diagram (see the example given in Figure 5.16).

REQ_0001 The behavior of a route must respect the following scheme:

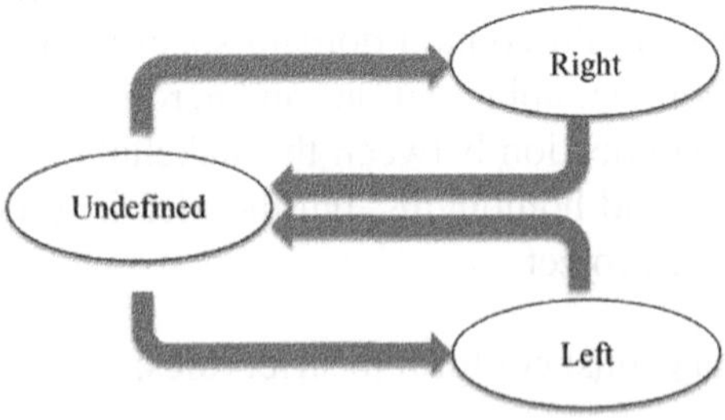

Figure 5.16. *Textual requirement supplemented with a diagram*

Formalization may go as far as the implementation of mathematical equations and/or construction of a so-called "formal model". By way of example, Figure 5.17 shows a B-model which is a formalization of the property P1: there must be no danger of collision. The property can be expressed, mathematically, in the following form:

$$\forall\{t_1,t_2\}\in[T], \text{so } D_{t_1}\cap D_{t_2}=\varphi, \text{if } t_1\neq t_2$$

```
MACHINE
    Example
SETS
        TRAIN
    ;   POSITION
VARIABLES
        Domain
INVARIANT
            Domain: TRAIN - -> POW(POSITION)
    &       !(t1,t2). ((t1, t2 : TRAIN*TRAIN) ⇒ (Domain(t1) ∧ Domain(t2) = {}))
INITIALISATION
    Domain :(     Domain : TRAIN - -> POW(POSITION)
                  & !(t1,t2 : TRAIN*TRAIN) ⇒ (Domain(t1) ∧ Domain(t2) = {}))
    )
END
```

Figure 5.17. *B-model of the property of non-collision*

5.3.3.4. *Formulation by use of a template*

The formulation template is an effective tool for the description of a requirement. The use of a requirement template will help to prevent the frequent mistakes that may occur during the description of the requirements in natural language.

The template is an excellent pedagogical tool, which is particularly effective during the learning stage. It is a guide to the syntactic structure of a unitary requirement, which is very helpful in the description of the functional requirements. An example of a template is presented in Figure 5.18.

By combining the glossary and the template, we shall be able to reduce the ambiguities introduced by natural language.

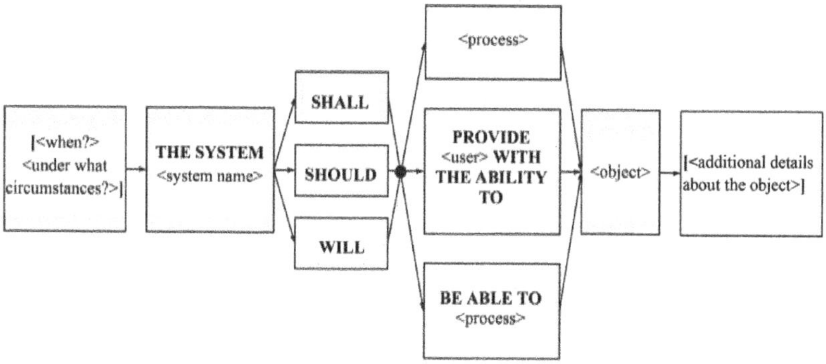

Figure 5.18. *Example of a template for the formulation of a requirement*[11]

5.3.3.5. *Modeling*

In order to work on the correctness and completeness of the requirements, it is interesting to create one or more models. A model is an abstract description of a system and/or a process. As it is a simplified representation, there is a reduction in the complexity. Hence, the model is easily manipulatable, and it enables us to reason and/or make a certain number of checks. The more accurate the model is, the closer the results obtained will be to those which will be observed with the final system.

Thus, the model can serve as a tool to aid communication between the different stakeholders.

A good model should focus on the problem, but it may go so far as to describe a solution, and in that case, it is possible to manage all or part of the final code. Therefore, a model may or may not be described as "executable".

11 Source translated from [POH 10].

More generally, it is possible to have models devoted to every step in the process of realization of the system. We then speak of Model-Based Development (MBD).

Modeling must be used in addition to textual description of the requirements: we must be able to express the needs in order to model them. The use of a model without a textual requirement is tantamount to discovering our requirements without having expressed them, and the issue arises of how to check whether the model corresponds to the requirement or not.

A model is usually composed of two mutually-complementary parts:

– a static model describing the entities constituting the system, and the states which may be associated with it;

– a dynamic model, describing the possible changes of state.

Chapter 8 of this book will be devoted to modeling.

5.3.4. *Requirements validation*

The requirements specification which has been produced enables us to identify the users' needs, and on that basis, the client can then prepare their battery of acceptance tests to perform on the system.

The purpose of the acceptance tests is to demonstrate that the system produced does indeed conform to the requirements. Therefore, the acceptance tests must cover all of the user requirements.

5.4. Requirements realization

5.4.1. *Process*

Figure 5.1 shows that once the requirements specification has been completed, it is possible to move on to the second phase, which is the realization of those requirements. The requirements realization phase must follow a cycle which should be capable of demonstrating that the final product meets the requirements.

It should be noted that we speak of the *realization* of an application, rather than the *development* of an application. The realization of an application encompasses the activities of development but also the activities of verification, validation, production, installation and maintenance of the application (see Chapter 4).

5.4.2. *Verification*

5.4.2.1. *Objective of verification*

The verification of a phase involves verifying the implementation of the quality requirements (application of the procedures, respect of the prescribed formats, etc.), the application of the processes (respect of the plans, respect of the organization, etc.), the correctness of the activities and the proper integration of the safety requirements.

For each phase in the realization process, it is necessary to conduct a phase of verification of the requirements, with three objectives – to ensure that:

– the initial requirement has been taken into account: traceability must be established between the upper-level requirements and the initial requirements. This traceability must be verified: the links exist and they are explicitly demonstrated;

– the set of requirements forms a correct whole: it is necessary to demonstrate that the requirements are comprehensible, unambiguous, verifiable, feasible, etc. We must also demonstrate that the set of requirements is complete and coherent (no conflict);

– no untraceable element has been introduced: the purpose of this verification is to check that all the requirements being formulated are traceable to a requirement of the next level. Very frequently, we see the emergence of so-called design and/or architectural requirements which actually have no link with the next level up. An analysis of these requirements shows that, for the most part, they represent things unsaid, and are very rarely true design requirements. It is also necessary to define what a design/architectural requirement is.

In the first and third points, the traceability study (analysis of each row of the traceability matrices) should show that all the requirements have been implemented, but also that everything that has been implemented is actually required.

Software Architecture Specification	Preliminary Design Specification
SAS_REQ_1	PDS_REQ_11, PDS_REQ_12, PDS_REQ_13
SAS_REQ_2	
SAS_REQ_3	DCP_REQ_11

Table 5.7. *Example of traceability matrix between an architecture and a design*

5.4.2.2. *Documentary review*

Verification is then performed by way of a quick "walkthrough" or a design review. The design review (see Figure 5.12) is a documentary check which must have an objective. This objective can be formalized in the form of a "checklist".

This checklist (see Table 6.8, for example) must define the checkpoints. These points are linked to knowledge of the types of faults which may be introduced during the activity which has led to the production of the documents under review.

Point	Rule	State OK/KO	Comment
R_1	All the parent requirements must be traced or a justification must be given		
R_2	All the requirements in the document must be traced to at least one parent requirement, or else a justification must be given		
R_3	All the interfaces identified must participate in at least one requirement		
R_4	All the states of the system must participate in at least one requirement		
R_5	Is each requirement atomic (no need to read through a set of requirements in order to understand the need)?		
R_6	Is each requirement verifiable?		
…	…		

Table 5.8. *Example of checklist regarding requirements*

5.4.3. *Traceability*

Figure 5.19 illustrates how the client's recommendations can be distributed throughout the system, and how the process can continue even to the level of the software- and hardware elements. In the level n_i verification phase, it must be demonstrated that the requirements on this level correlate to the next level up: n_{i-1}. This correlation check is performed by implementing an action of traceability.

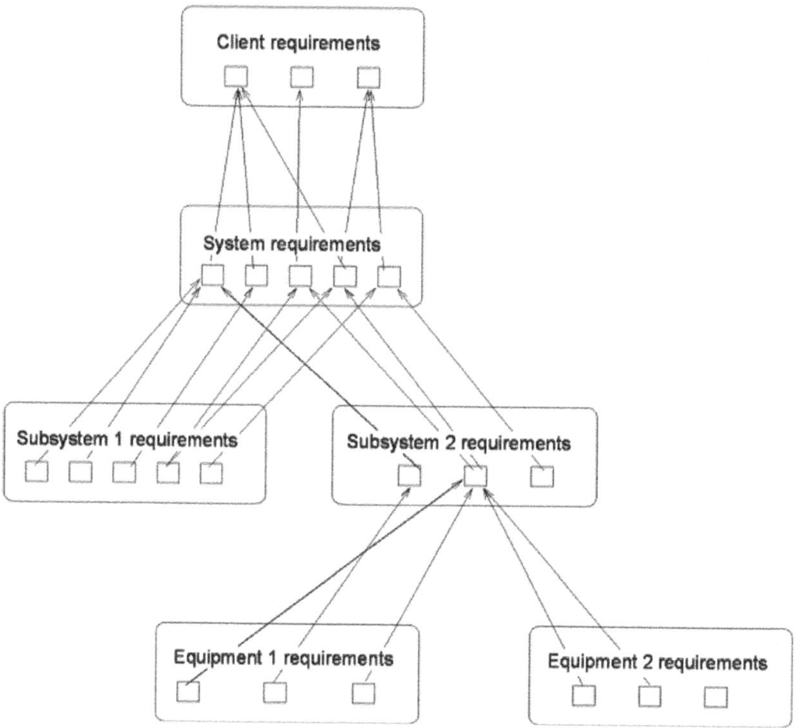

Figure 5.19. *Partial traceability between the client requirements and the equipment-related requirements*

The implementation of traceability (see Definition 5.2) involves the definition of at least one link between two objects. In the context of requirements, the traceability links must be able to demonstrate that a requirement on level n_i is connected to a requirement on the previous level, n_{i-1}. The inverse link demonstrates that no requirements have been overlook during the realization process.

DEFINITION 5.2 (Traceability).– *Traceability consists of establishing a link between two objects.*

The traceability must be bidirectional, and we must be able to establish:

– vertical traceability: tracing a requirement from its highest level to its lowest, with the objective being to show that whatever the level, all the requirements are related to a need;

– horizontal traceability: tracing a requirement through the phases of the process of development and realization. The objective is to show that a requirement has been satisfied.

In requirements management, there are three kinds of traceability:

– upward traceability of requirements;

– downward traceability of requirements;

– inter-requirement traceability.

When traceability has to be implemented for the requirements, it is useful to evaluate and define:

– the level of detail of the traceability connection, which corresponds to the detail of the elements which it connects. Does it trace single requirements? Sets of requirements? Use cases? Whole documents?

– the semantics of the traceability connection: "derives from", "satisfies", "realizes", "verifies", "uses", "depends on", etc.

There are several ways to implement traceability:

– by a textual reference;

– by a hyperlink;

– by a reference internal to the tool.

As is illustrated by Figure 5.19, there are a number of basic transformations of requirements. Of these, two cases are particularly interesting: the addition and the abandonment of a requirement. In one or other of these cases, it is absolutely essential to attach a justification to the requirement.

As we have seen, the aim of traceability is to define links between the different requirements. As Figure 5.21 shows, the basic link connects two requirements which belong to two different, consecutive levels.

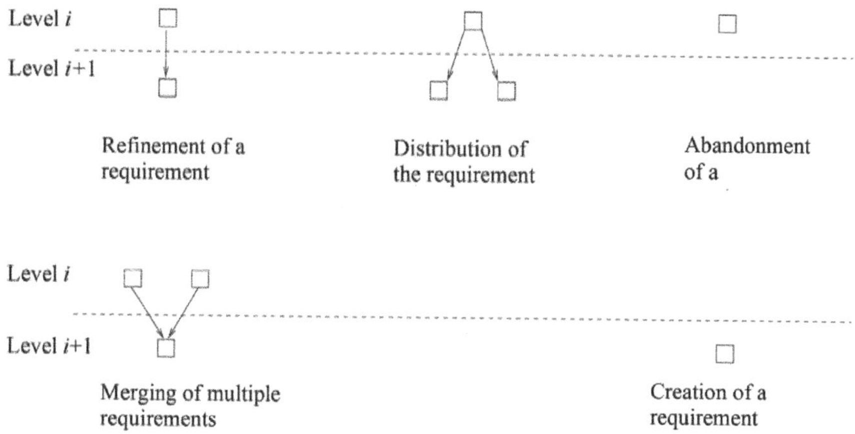

Figure 5.20. *Basic traceability of the requirements*

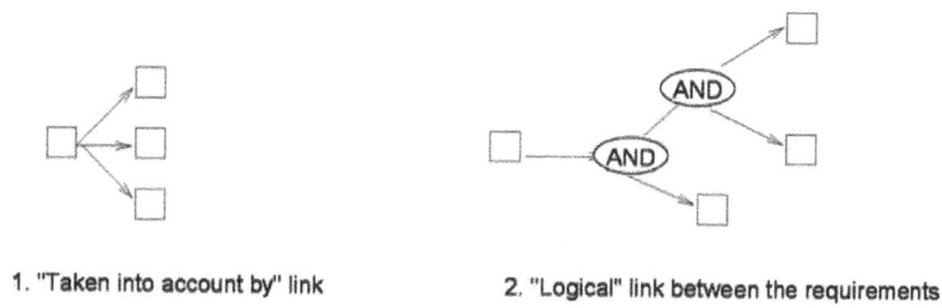

Figure 5.21. *Link between the requirements*

The requirements are then associated with system functionalities. A functionality is a behavior which is expected of the system. The term "functionality" is used because at system level, there may not necessarily be a service in the sense of a "function", but rather an overall service (which is the result of a set of actions). Naturally, this process must be divided between the subsystems, the devices, the hardware aspects and the software aspects.

Figure 5.22 shows a traceability matrix which illustrates the link between the requirements and the functions, but also between the requirements and the secondary requirements derived from the former.

Item 1 in Figure 5.22 shows the functional link and the association between the requirements on the two levels. The function Fct1 is split into three sub-functions. As item 2 shows, the requirement Ex2 has been divided into two (Ex21 and Ex22) which are associated with two sub-functions (Fct12 and Fct13) of function Fct1.

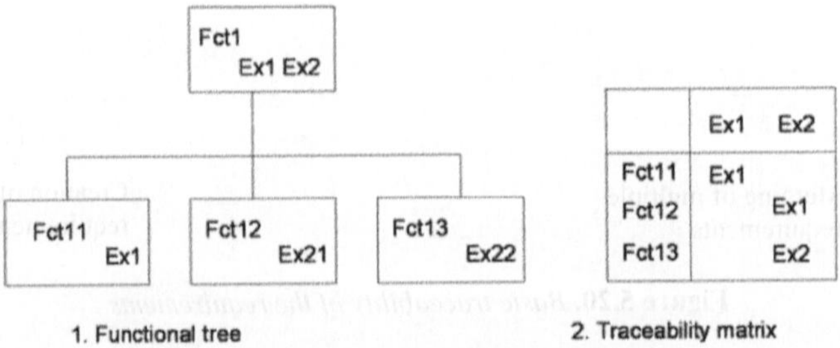

Figure 5.22. *Requirements traceability matrix*

5.4.4. *Change management*

5.4.4.1. *Introduction*

Figure 5.1 illustrates a process wherein the requirements realization phase cannot be launched until all the requirements characterizing the need have been acquired and grasped. Indeed, in case of a premature launch of the requirements realization phase, any evolution of the need-related requirements may have an impact on costs and time (see [STA 94] and [STA 01]).

There are two types of projects:

– for certain projects, the requirements acquisition phase is performed by the person responsible for lodging the order, and is part of a very controlled, well-managed process. In this context, the referential framework at the end of the requirements acquisition phase is relatively stable, and the evolution of this framework will give rise to an evolution of the contract;

– for the second category of project, the acquisition phase forms part of the project, where the response to a call for tender has been made before completion of the acquisition phase. In such a case, the set of requirements is not known, and may evolve for a certain period of time. It is then necessary to have a process in place for managing the changes.

In addition to these two categories of projects, we must also modulate our ideas with the fact that the design of certain products needs to take account of regulatory and normative objectives. Indeed, any regulation in force represents an additional set of requirements for which to cater, but this has an impact on the understanding of the product, meaning the requirements realization process is more likely to be delayed until the acquisition phase has been completely finalized.

5.4.4.2. *Nothing is perfect*

As we have seen in the previous sections, the definition of the requirements is a process which, on the basis of non-formalized elements, seeks to identify the requirements which the system must respect. With this goal in mind, the process must necessarily be iterative: it must be accepted that the requirements are subject to change. This means that the initial requirements may be improved during the design of the application.

Such changes may be induced by a difficulty in interpretation during the refinement and/or attribution of the requirements to the components, by modification requests, by a difficulty (or impossibility) to perform the tests, etc.

In order to be able to handle this iterative process, a specific attribute defining the state of a requirement may be added. This attribute could, for example, take the values: tabled, under analysis, accepted.

5.4.4.3. *Accepting change*

Today, clients are no longer prepared to accept inflexibility; they want to be free to ask for functional changes to be made to a project during the course of its development. Hence, the designers of new products must have a positive attitude toward change.

The development of so-called *smart products* must respond to a market dynamic where the clients' requirements tend to change very rapidly in an environment of fierce competition. In this frantic rat-race, innovation must be at the heart of the definition and the development of new products, and it is a major advantage to be the first to launch a revolutionary product on the market – the so-called *killer product.*

The evolution of the regulations, which impose increasingly-stringent constraints on the functionalities and design of these products, also contributes to this dynamic of change, but to a lesser extent.

As a product designer, it is essential to be able to act swiftly when faced with a new requirement or a changed understanding of the client's needs during the development process. This ability to accept and deal with change is a crucial pillar in agility. Thus, we need to be able to devise new engineering processes in order to respond to a "natural" evolution of the requirements during the course of the project.

These "agile" processes must be anticipated and defined as early as possible, and must not be the result of an *ad hoc* process. In this context, requirements management is a central process, because it is this process which will ensure that the client's requirements are fulfilled and that there is traceability between those requirements and the end product finally delivered. We need to think of the requirements as being evolutive artifacts, which are always crucial to the success of the projects.

5.4.4.4. *Impact of change*

Analysis of a change is performed by way of an impact analysis (see Definition 5.3) and a non-regression analysis (Definition 5.4). In certain cases, non-regression is said to be total. In such cases, it is then necessary to re-execute all the tests in one or all of the phases. The purpose of non-regression analysis is to minimize the cost of engineering a new version.

DEFINITION 5.3 (Impact analysis).– *Impact analysis of an anomaly consists of identifying the modifications that need to be made on the later phases of the realization (impact on the documents, impact on the code, impact on the description and implementation of the tests).*

DEFINITION 5.4 (Non-regression analysis).– *Non-regression analysis consists of determining a set of tests to demonstrate that the modification that has been made has no effect on the rest of the software application*[12].

Requirements management is a tool that can be used to manage any evolutions made after implementation of the original system. Indeed, it is possible to define the impact cones relating to a requirement.

As is illustrated by Figure 5.23, on the basis of the modified requirements, it is possible to perform an impact analysis. In order to do so, we extract the cone whose origin is the modified requirement and which works toward the code. This code enables us to identify all the requirements of the lower levels which could, potentially, be impacted by the modification, and the test cases associated therewith.

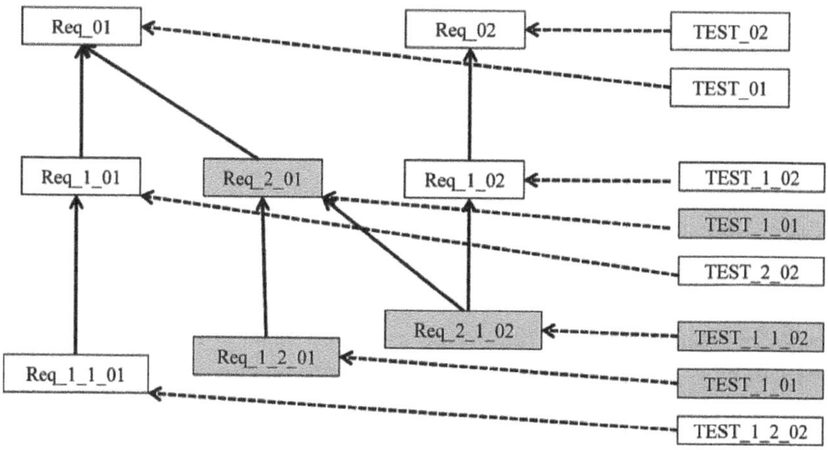

Figure 5.23. *Impact analysis*

The second analysis (see Figure 5.24) involves performing non-regression analysis. Non-regression analysis entails constructing a cone which begins (at its vertex) with the modified requirement, and works back toward the requirements in the initial specification. Based on the requirements cone, we can thus select the non-regression tests to repeat in order to show that the modification does not impact on the system requirements.

12 It should be noted that non-regression testing can be performed on the software application or on a larger element, such as a device, a subsystem and/or even a system.

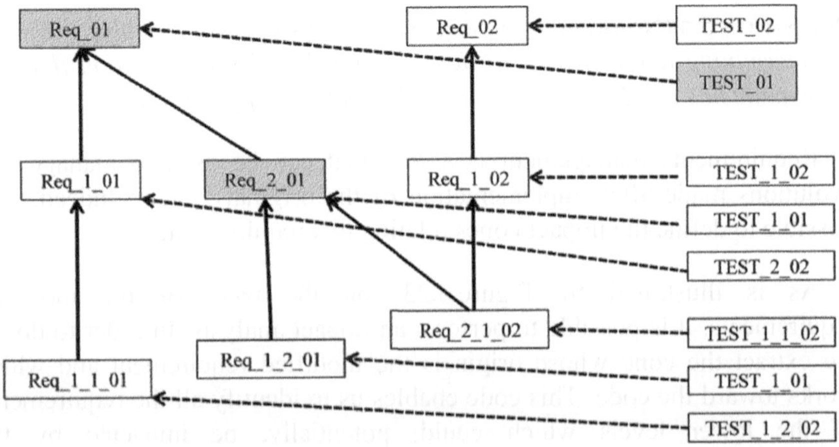

Figure 5.24. *Non-regression analysis*

5.5. Requirements management

5.5.1. *Activities*

Requirements are used to form the link between the documents (specification lists, design files, encoding files) but also as testing objectives. The various categories of tests (unit tests, software/software integration tests, hardware/software integration tests, functional tests and income tests) can be viewed in relation with the requirements on the basis of the attribute "type of verification". Thus, we obtain a requirements-management process which involves forming the link between the initial needs, the choices made during the realization, and the validation phases (see Figure 5.25).

The phase of verification consists of verifying the fulfillment of the requirements and the coherence of the links that have been established. From the standpoint of the project, management of the fulfillment of the requirements enables us to quantify the work that has been done and the work that remains to be done.

Management of the evolution of the requirements and analysis of their impacts on the linked requirements and on the products are the key points of requirements management. It must be noted that the true difficulty of requirements management lies in management of the evolutions.

In summary, we can state that requirements management requires simple mechanisms to be put in place, such as:

– the introduction of identifier management;

– the description of the requirements;

– the definition of a traceability matrix.

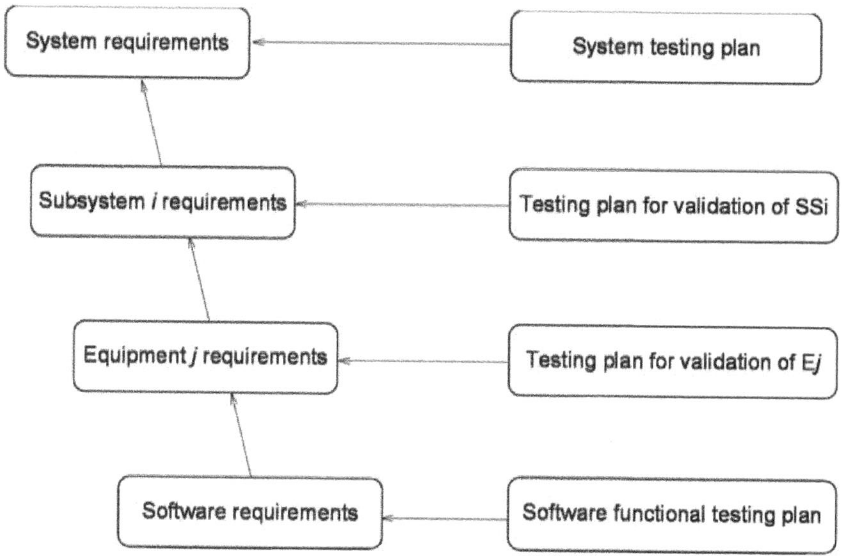

Figure 5.25. *Links between the activities*

5.5.2. *Two approaches*

Today, requirements management is enacted by way of two approaches:

– database-centered approach;

– document-centered approach.

5.5.2.1. *Database-centered approach*

The first approach, which is called the database-centered approach, involves compiling all of the requirements (and the elements associated therewith) in a database, and establishing all of the links within that database.

The use of this approach involves being capable of importing all the elements (documents, models, source files, test scenarios, test results, etc.) into the database. If this is done, it becomes possible to produce the traceability matrices and think about the links existing between the requirements (impact analysis, evolution management, etc.).

However, centralization of the information in a single database is a disadvantage, for various reasons:

– the size of the database depends on the size of the documents being handled;

– in case the source document evolves, it is necessary to re-import it into the database. However, it is usually rather difficult to automate the reintegration of data relating to the requirements;

– etc.

5.5.2.2. *Document-centered approach*

The second approach, which is known as document-centered, consists of going through the documents (documents, models, source files, test scenarios, test results, etc.) and adding information regarding the links between the objects. That information comes in the form of tags, attributes and links.

The use of this approach requires us to be able to add data and also to harvest them from the requests submitted.

The advantage of this approach is that it does not alter the process that has been established for the conduction of the project, and is limited to the embellishment of the existing documents.

5.5.3. *Implementation of tools*

In [STA 01], mention is made of the fact that the establishment of a requirements-management environment is the best way of producing a major impact on the success of a project. The definition of a minimal set of requirements provides us with a base that can be managed, and the tool then becomes a vector of communication between the various teams involved.

A text-editor-type tool[13] may suffice in this case (with tables, identifier management, links between the documents, etc.) to handle one or several documents. In complex systems, the number of documents and number of steps (see Figure 5.25) create a need for a tool-based requirements-management process.

A tool is not sufficient to manage the requirement, it is necessary to define a management process. In another way, it is preferable to define a management process and to select the appropriate tools to implement it afterwards.

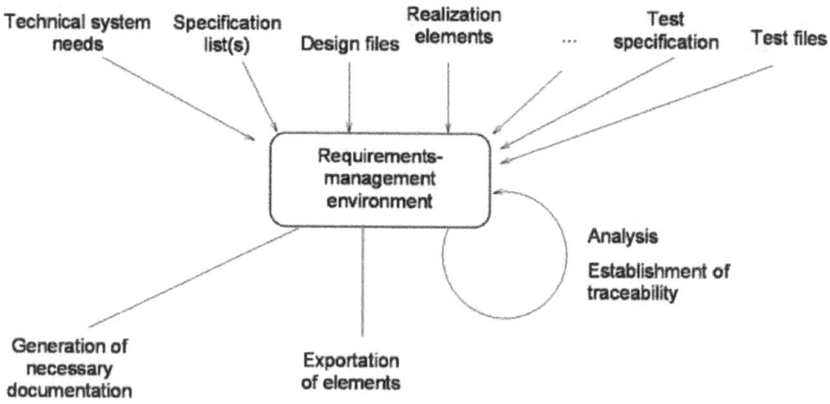

Figure 5.26. *Outline of a requirements-management environment*

[CHO 01] presents the approach known as Common Airbus Requirements Engineering (CARE), which was designed to aid in the development of the AIRBUS A380. This approach is based on the definition of a global methodology, which is inspired by the EIA632 standard [EIA 98, EIA 03] and is equipped with a devoted toolkit.

Requirements management is supported by tools which facilitate the acquisition of the requirements, the establishment of traceability measures,

13 As part of the twofold validation implemented by the RATP for the SAET-METEOR equipping Line 14 of the Paris Metro, the text-processing tool INTERLEAF was used to define and manage all the requirements (from system level to the three software programs written in the formal language "B").

reporting and the generation of documentation. This list includes, for example, DOORS (distributed by IBM), RTM (Integrated Chipware Inc.), Requisite Pro (Rational) and Reqtify (created by Dassault for traceability matrix generation). One of the main issues associated with the use of these tools is their integration into the trade processes used by individual companies.

5.6. Conclusion

Requirements management is an entirely separate process. For that reason, it is essential to produce a requirements-management plan, the goal of which is to describe:

– the organization put in place to manage requirements on the project during all the phases (from the requirement acquisition phase to the decommissioning phase);

– the concept of requirement: syntax, principle of drafting, list of attributes, etc.;

– the resources mobilized to manage the requirements: tools and processes;

– the principles of document management and production;

– the process of evolution management (identification of changes, impact analysis, non-regression testing);

– etc.

For readers wishing to know more about requirements, we recommend consulting [BOU 14a, HUL 05, POH 10].

6

Data Preparation

6.1. Introduction

In recent years, the process of installation of a software-based system has undergone a change, caused by the need to establish product lines. Indeed, no longer is a product designed solely for one use; rather, it must be able to be used in various contexts. This issue is common to the various domains and the standards associated therewith [BAU 10].

When a system is dependent on a dataset to define and realize its behavior, we say that it is a parameter-based system. There may be various types of parameters: data describing the context of the system's use (topology, velocity curve, stopping points, etc.), data governing the services (maximum number of objects, weight, etc.) and service activation/inhibition data where some services such as Electronic Stability Program (ESP) in the car are present and can be activated when the user requests.

In parameter-based systems, the safety of the complex, operationally-safe critical system is governed not only by the validation of the software application but also the validation of the data handled by that software application.

The management of parameter-based software applications is described in section 8 of CENELEC 50128:2011 [CEN 11a], which we shall present in this chapter.

6.2. Recap

The railway systems in charge of signaling have the characteristic of handling data describing the topology of the line that they manage (see [GEO 90], for example). These topological data are said to be invariant, because they do not change unless work has been carried out on the line. The topological data include lights, routes, track circuits, platforms, the description of the line, etc. Topological data are not the only invariant data that can be used by a railway system; there are also data pertaining to the features of the train: its weight, its length, its maximum speed, maximum acceleration, braking characteristics, door characteristics, length of alarm on doors closing, etc.

Signaling systems, but many other railway systems as well, can thus be constructed on the basis of a generic application (applicable to different sites and different users) and a configuration process [BOU 00, BOU 03, BOU 06] used to obtain a specific application (dedicated to a single site, one use, a lone user).

The CENELEC 50128:2011 standard (see Figure 2.8) covers both aspects: the development of a generic software application (see Chapter 7 of the standard) and the development of a process to manage the parameters of a generic application (see Chapter 8 of the standard). The aim of the current chapter is to present the issue, the concepts associated therefore and what is expected of data preparation, through an analysis of CENELEC 50128:2011.

6.3. Issue

Global Positioning System (GPS) is a tool that is very widely used in daily life today, and is an example of a system where data are ubiquitous. For GPS, it is difficult to imagine there being any direct impact on safety, although certainly some surprising situations could arise (e.g. lack of any road, a non-optimal route or one which disobeys signage, etc., as illustrated by Figure 6.1), but as the driver is still ultimately responsible for the decisions, the GPS is merely a driving aid. However, new uses for GPS, such as the monitoring of train positions, could have an impact on safety-related recommendations such as measuring of the position, accuracy, etc.

As we shall see later on, the railway domain is one example where data are ubiquitous in the realization of the system, but it is not the only one. The nuclear, automotive and space domains are other areas where data management can impact on safety.

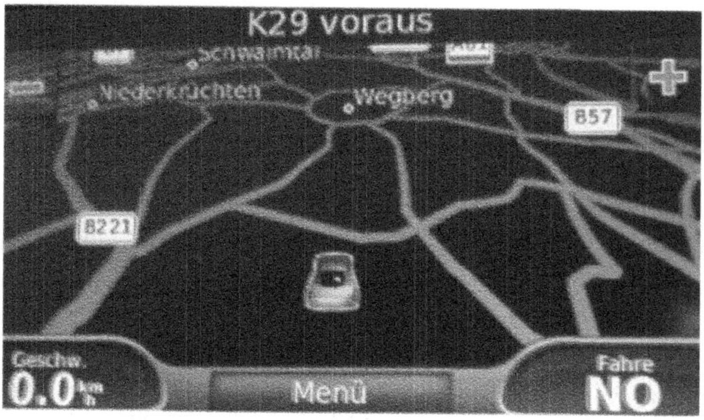

Figure 6.1. *Example of issues relating to data*

For example, in the nuclear domain, the concept of parameter data is essential. Thus, the systems SPIN (For "SeParation & INcineration" – a French federally-mandated program) and US (*Unité de Surveillance*, or Monitor unit) for N4-class[1] reactors generate a vast quantity of data (more than 10,000 units). Some of these data are updated during the operation of the reactor – notably to take account of the depletion of fuel. At least two data-related incidents are identified on the Website of the ASN (*Autorité de Sûreté Nucléaire*, or Nuclear Safety Authority – www.asn.fr). Following the incidents on 14 February 2000 and 4 April 2000, the analysis shows that it was mistake in parameter-setting which led to an underestimation of the power released by the fuel in a given situation, leading to automatic shutdown of the reactor. Note that it was the parameters of the SPIN and US that were at fault. These errors were due to the poor quality of the documentation used in programming the systems.

For the nuclear domain, section 5.2 of the IEC 60880 standard [IEC 06] introduces the concept of configuration data as being data that are necessary for the adaptation of applicative software to the inputs/outputs and the services required by the installation.

1 The N4 classification applies to pressurized water reactors with 1450MWe nominal power, operated by EDF.

Parameter data are divided into two broad categories:

– data which are not intended to be modified online by operators, and which are subject to the same requirements as the rest of the software;

– parameters, i.e. data which can be modified by operators during the operation of the installation (e.g. alarm thresholds, checkpoints, instrument calibration data), and which are subject to specific requirements.

In the spatial domain, the preparation for launch of an Ariane 5 rocket ([BOU 10] Chapter 7) requires the production of the data relating to a trajectory. These data must take account of the characteristics of the launcher, the trajectory identified and the weather conditions. There are three types of such data: the mission data (mass, intended orbit parameters, etc.), family data (type of nose on the rocket, etc.), configuration data and data dependent on the launcher (sensor calibration, etc.). However, before it is possible to produce the data, it is necessary to carry out studies and simulations. The data (in tabulated form) are produced in the form of ADA packaging [ANS 83, ISO 95, BAR 14], and can be exploited by the flight program.

6.4. Data-parameter-based system

6.4.1. *Introduction*

The command/control systems being developed today are increasingly large and cumbersome; they integrate the concept of communication and consume data. These statistical data may be of two types:

– fixed data, said to be "invariant", which characterize the environment (topology, trajectory, etc.), the system (number of objects, limit, etc.) and characteristics (weight, length, velocity, etc.);

– data which evolve with the state of the system.

DEFINITION 6.1 (Specific application).– *A specific application is a generic application with which a set of parameters has been associated. This specific application can only be used for one installation.*

It should be noted that the concept of a specific application covers both the software and hardware aspects.

DEFINITION 6.2 (Generic application).– *A generic application comprises an execution platform (called a generic product) and a software application. That software application is defined on the basis of a set of parameters which must be instantiated as a function of the final use (depending on the site, depending on the services needing to be provided, depending on the technical characteristics techniques, etc.).*

As Figure 6.2 illustrates, the generic application (see Chapter 7) is therefore the combination of a generic software program and an execution platform. The execution platform is called the generic product, and covers the hardware aspect and the software aspects (operating system, firmware, middleware for communications management, etc.).

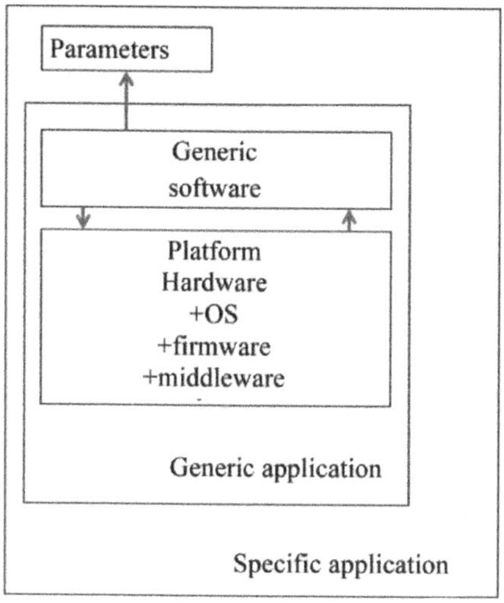

Figure 6.2. *Specific application versus generic application*

DEFINITION 6.3 (Generic product).– *A generic product comprises a set of hardware elements and a set of software programs (operating system, firmware, middleware, etc.). The aim of a generic product is to facilitate the execution of a software application. A generic product can be used for the realization of a variety of systems.*

As is indicated by Definition 6.3, a generic product is designed to execute different types of software application – the generic aspect means that it does not need to be specialized; the concept of a product is linked to the fact that this set can be certified independently of the field of use. Such is the case, for example, of programmable logic controllers (PLCs).

The generic product is, in general, developed independently from a specific project. See, for example, the different PLCs already certified SIL3 in conformance with IEC 61508 ([IEC 08]) or the internal generic products developed by railway industrial actors based on 2oo3 architecture or coded monoprocessors (see [BOU 10a], Chapter 2).

The invariant data are known to the generation of the system and define a configuration for the system. Thus, it is possible to generically define a system and implement it by way of parameters on another site.

For example, from the description of a metro line, it is possible to deduct a dataset common to all the devices making up the system. Those data depend on the topology of the line, the characteristics of the trains and the technological choices (response time, processor computation time, etc.). In the world of railway systems, they are called "topological invariants" [GEO 90].

A metro line (see the fictitious example given in Figure 6.3) is made up of tracks carrying train traffic. Those tracks are constituted by track circuits (TCs). Each track circuit is a portion of track which is able to detect the presence of trains in that particular area. The track circuits are generally connected to two other track circuits, but in cases where they are connected to three other track circuits, they must be associated with a route, which is a device that enables a user to choose the path to take.

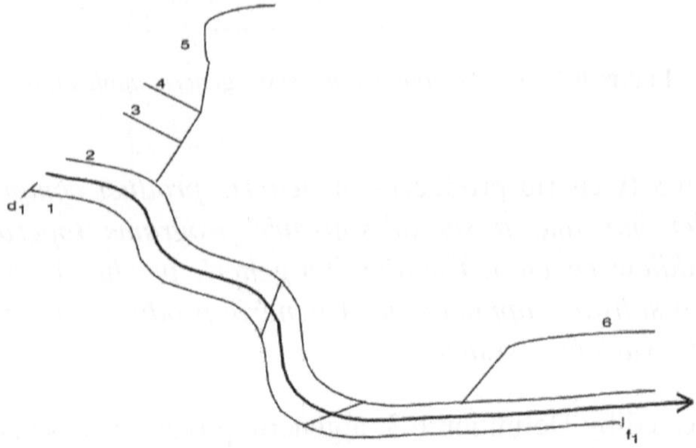

Figure 6.3. *Example of topology*

6.4.2. *Characterization of data*

A specific application (see Definition 6.1) is therefore devoted to a specific use. This application uses data of different types:

– service activation/inhibition data;

– fixed data which do not evolve over time. Such data describe the system's characteristics (topology, velocity curve, stopping points, etc.);

– data which evolve over time, with the evolution time being the cycle or a longer duration.

Data whose values change regularly (in the railway domain, these data are called "variants") may be:

– input that the system acquires regularly;

– data that the system calculates, such as the global or local variables and the system outputs.

Definition 6.4 (Configuration data).– *Configuration data are parameters whose value does not change during the execution of the software application.*

As is indicated by Definition 6.4, the parameter data (in the railway field, we speak of "invariants") are the first type of data. The parameter data generally include two families of data: those defining the technical characteristics (velocity, weight, size, number, etc.) and those defining the use context (the topology, the velocity curves, the stopping points, etc.).

The parameter data characterize the implementation of a so-called generic software application to a specific case. Note that there are two types of configuration data (see Figure 6.4): the so-called calibration data, which are deemed to be invariant because they change extremely slowly, and the configuration data which, for their part, actually are truly invariant.

DEFINITION 6.5 (Calibration data).– *Calibration data are data which evolve over a "long" period of time.*

As Definition 6.5 indicates, the calibration data are linked to the fact that the system's characteristics evolve over time (decrease of fissile material in a nuclear power plant, evolution of the characteristics of engines because of wear and tear, etc.). Railway systems use some calibration data, for example, to confirm the wheel diameter.

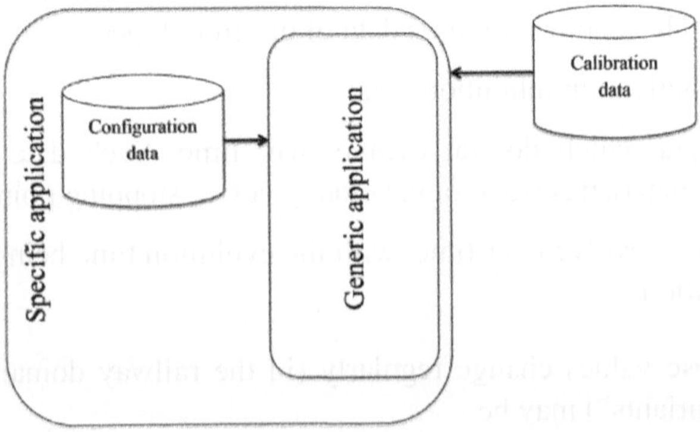

Figure 6.4. *Configuration data versus calibration data*

6.4.3. *Service inhibition*

The configuration data may be associated with real data for the system, but in certain cases they are associated with services – as is the case, for example, with service activation/inhibition data. Such data (see Figure 6.5) are used to determine whether or not the execution of the services can be allowed.

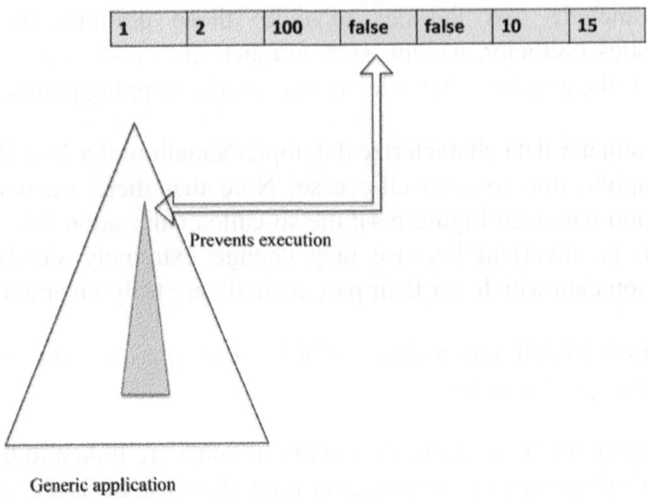

Figure 6.5. *Data controlling the activation of a segment of code*

The use of inhibition/activation data is therefore associated with the notion of dead code (i.e. code which cannot be executed in a real-world execution). Thus, it is necessary to obtain a complete validation of the application with and without the data-controlled functions to ensure that a runtime error (such as a memory anomaly, a processor anomaly, inadequate point management, etc.) cannot lead the software application into an uncontrolled situation.

This type of data can be used to put in place an activity of capitalization, and therefore establish a process of reuse of services that have already been validated. However, it is not compatible with the *a priori* creation of unnecessary functions for future use. Whilst it is true that this approach could help reduce the cost of a future project by reusing the existing functions, it could also cause costs to spiral out of control if functions are developed *a priori* that are not used on the project.

The inhibition/activation data are useful in defining a product line. Figure 6.6 presents a process of realization of an application based on the concept of a product line. The generic product is not an entirely separate application: it is necessary to put in place a stage of specialization, whereby we define the common boundary between the common core and the generic application to be realized; using that knowledge, it is possible to construct the generic application on a shared base.

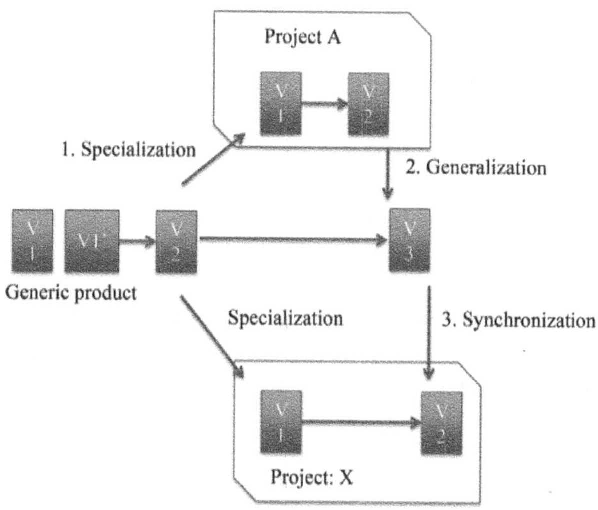

Figure 6.6. *Product line*

This process is associated with two activities:

– the first, which we can call "generalization", involves harvesting the behaviors of the generic application which have been validated in the common core. We can then use them "as are" in a subsequent software-development project;

– the second activity is called synchronization, and is linked to the fact of updating a generic application in relation to the common core; indeed, that common core may have evolved (with error correction, generalization, etc.).

Here, we shall not further describe the concept of a product line, although it remains a challenge to control the cost of the realization of software applications. CENELEC 50128:2011 [CEN 11a] does not identify this issue.

6.4.4. *Overview*

In this section, we have laid down the specific vocabulary that is used to speak of parameter-based software applications. In the railway domain, the concepts of a generic product, a generic application and a specific application may facilitate re-use, but above all they are essential in dividing safety research.

CENELEC 50126 [CEN 99][2] – which describes the management of the safety-related processes – is applicable from the level of the whole system to that of the generic product, and CENELEC 50129 [CEN 03] – which focuses on the formalization of the safety demonstration – introduces the concept of a safety case and the possibility to compile safety cases for the generic product, for the generic application and/or for the specific application.

Finally, it is possible, in the railway domain, to realize systems on the basis of a generic product, a generic application and/or a specific application, and only configuration data are used to set the parameters.

2 See Chapter 2 for an introduction to the CENELEC standards, and more specifically to CENELEC 50129 and the concept of a safety case.

6.5. From the system to the software

6.5.1. *Need*

Parameter-based software applications require specific work to identify and produce the data to be handled. This activity is, for the moment, identified as an activity to be carried out during the phase of realization of the software application, but at present, there is a danger of constructing a system which will not work.

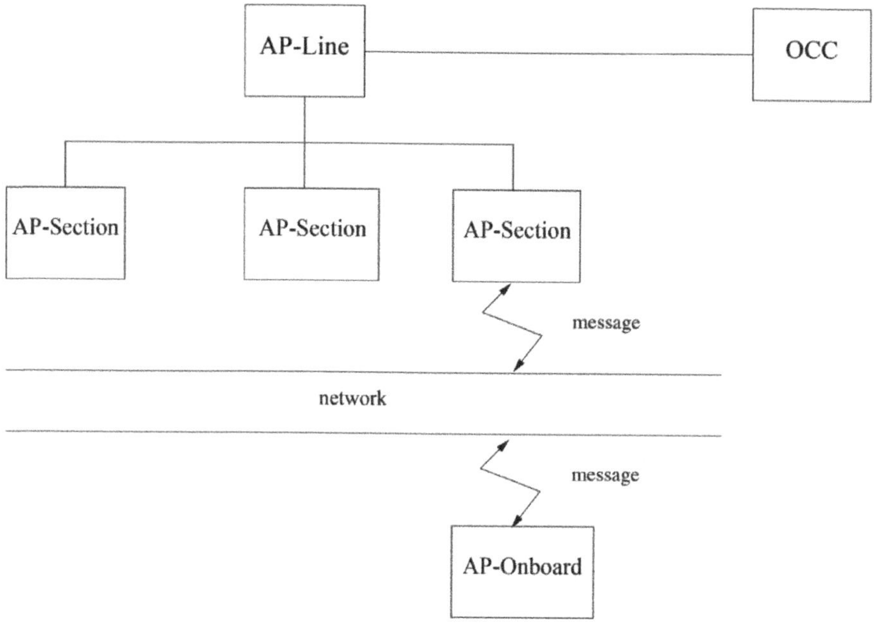

Figure 6.7. *View of a system*

Indeed, a railway system (for example, see Figure 6.7[3]) comprises many different kinds of equipment (OCC, AP-Line, AP-Section, AP-Onboard and transmission belt) and may have several software applications – either parameter-based or otherwise (see Figure 7.8).

3 The example presented here is connected to SAET-METEOR, which is the system employed on Line 14 of the Paris Metro. Further information can be found in Chapter 2 of [BOU 12].

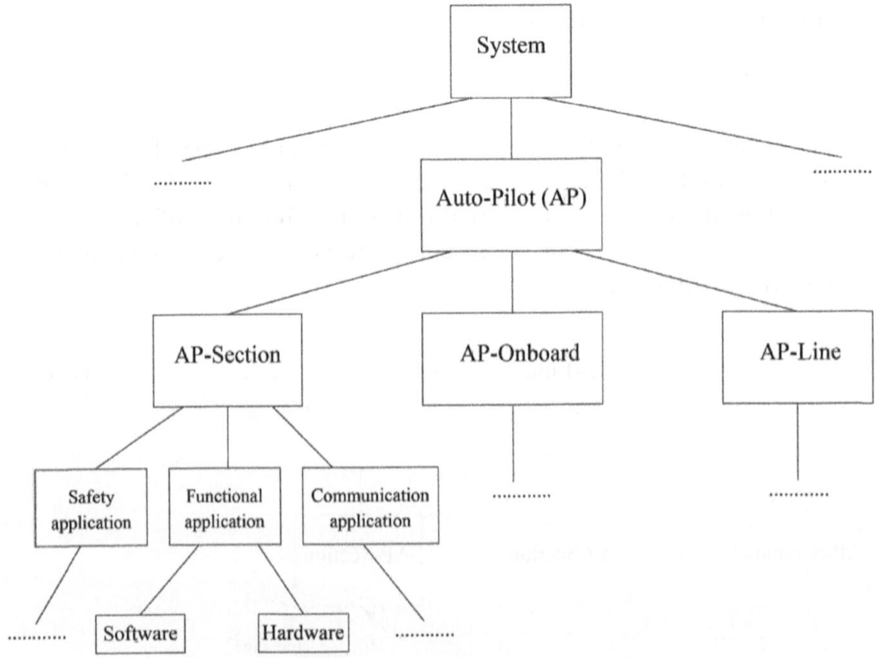

Figure 6.8. *From the system to the software*

In such an approach, the main difficulty relates to the coherence of the data between different devices. The data must (non-exhaustive list, but one which does offer a general idea of the issue):

– describe the same system: common source of input for the production of the data, identical version of the data, technical interpretation of the characteristics of the data, etc.;

– be coherent: compatible precision, same frame of reference of the units (km, etc.), similar frame of reference for orientation (where is the point of reference?, etc.), compatibility of the frames of reference (the switch from a *PK* orientation to a track marker-based orientation, etc. – see Figure 6.9), identical limits (the maximum speed of the train must be the same for all the devices, etc.);

– be correct: the same activity of validation must show that the data are correct and that the operation of the system is in line with what is required for all configurations.

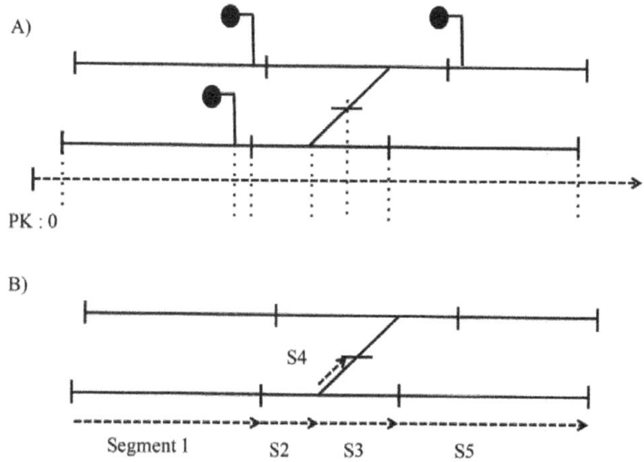

Figure 6.9. *Pinpointing of the objects: a) on the basis of a "kilometric point" (PK – European system), b) on the basis of a segment*

Figure 6.9 shows an example of topology, and demonstrates that it is possible to have different frames of reference by which to pinpoint the objects. Initially, the objects are located by using a measurement in relation to "kilometric points" (PK, roughly equivalent to "mile markers" – see case (a) in Figure 6.9), but at the equipment level, the tendency is to work with more precise local frames of reference, such as the segment (case (b) in Figure 6.9) or the track marker.

6.5.2. *What the CENELEC framework does not say*

The concept of data is essential for railway systems, but CENELEC 50126 [CEN 99], which describes the process of safety management throughout the lifecycle of the system, does not directly mention this concept. Meanwhile, CENELEC 50129 [CEN 03], which is intended to formalize safety demonstration, does indeed introduce the concept of generic products and generic applications, but does not indicate how the parameters should be managed, and what form they should take.

As Figure 6.10 illustrates, the data-preparation process to be implemented comprises three steps:

– step 1: identification of the data at system/subsystem/equipment level. This identification enables us to define the data and their characteristics;

– step 2: identification of the data consumed by the generic software. On the basis of the system data and the software requirements, during the design of the software, we must identify the data that are actually consumed and the characteristics of those data;

– step 3: definition of the data production process, on the basis of the system data and the data necessary to set the parameters of the generic application, a data preparation process must be defined. We must show that the data stemming from the data preparation process respect the desired properties. This data production process should enable us to demonstrate that the data conform to their SSIL objectives [BOU 99, BOU 00].

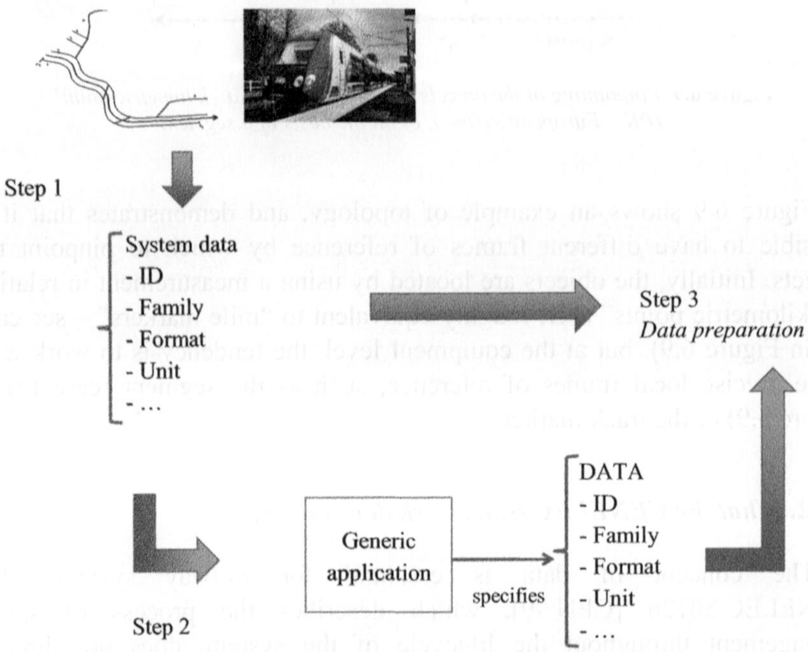

Figure 6.10. *A three-step process*

The data preparation process was identified in the previous version of CENELEC 50128 [CEN 01a], but it was impracticable. In the 2011 version of the standard, section 8 identifies the principles characterizing the data and the needs to carry out a data preparation process that respects CENELEC 50128:2011 [CEN 11a] and the objective SSIL.

6.6. Data preparation process

In this section, we shall present the data preparation process as stipulated by CENELEC 50128:2011, and more specifically, we shall explain what is expected because of section 8 of that standard.

6.6.1. *Context*

6.6.1.1. *Software assurance*

As was demonstrated by Figure 2.8, Clause 6 (which relates to software assurance) needs to be implemented both for the realization of the generic application and for the parameter setting. It is therefore necessary to have quality assurance in accordance with ISO 9001:2008 [ISO 08], skill management, a form of organization, configuration management, verification/validation, to qualify the tools and realize an evaluation.

Therefore, it is important to have a set of plans (SQAP, SCMP, VVP, SVaP, etc.) that describe the organization, processes and resources mobilized to realize the data preparation processes. These plans may or may not be specific; the data preparation may be part of a project or be an entirely separate project.

Software assurance was presented in Chapter 4 and is still applicable for the generic software, to the preparation of the principles of parameter setting and to specific parameter setting.

6.6.1.2. *Safety assurance management*

CENELEC 50128:2011 does not identify an activity of safety assurance, but it requires the implementation of CENELEC 50126 [CEN 00]. CENELEC 50126 defines a RAMS management cycle (see Figure 2.4) which is based on acquisition of the need when the system is decommissioned. For our purposes, though, CENELEC 50126 requires the safety team to identify the functional requirements pertaining to safety and the safety integrity level requirements, and to take account of all the requirements throughout the whole of the cycle.

CENELEC 50129 [CEN 03], which pertains to the safety case and is normally applicable to the signaling subsystem (in the next version of

CENELEC 50126, this section will be applicable to all parts of the system), introduces different levels of the safety case:

– the safety case for the generic product;

– the safety case for the generic application;

– the safety case for the specific application.

This structure conforms to Figure 6.2 and to Definitions 6.1, 6.2 and 6.3, given above.

Hence, the safety team must, in its safety assurance plan (SAP), define a safety demonstration strategy which takes account of that decomposition (Figure 6.2). A set of activities needs to be established with a view to showing that:

– the safety management requirements are properly served at every level;

– the generic product is indeed safe and it clearly identifies the exported constraints pertaining to its use;

– the generic application is able to conform to the constraints exported by the generic product, that it is safe and clearly identifies the exported constraints pertaining to its use;

– the specific application is able to conform to the constraints exported by the generic application, that it is safe and clearly identifies the exported constraints pertaining to its use;

– the specific application is consistent with the final installation.

6.6.2. *Presentation of section 8 of the CENELEC 50128:2011 standard*

6.6.2.1. *Lifecycle*

The structure given in section 7 of the CENELEC 50128:2011 standard adheres to the V-lifecycle, and for each phase, the standard identifies the inputs, outputs and activities. For section 8, the structure is different; indeed, the process of production of the parameter data is viewed as a whole. Section 8.2 identifies the inputs, section 8.3 the outputs and section 8.4 describes the cycle (see Figure 6.11) and the activities that need to be carried out in relation to that cycle.

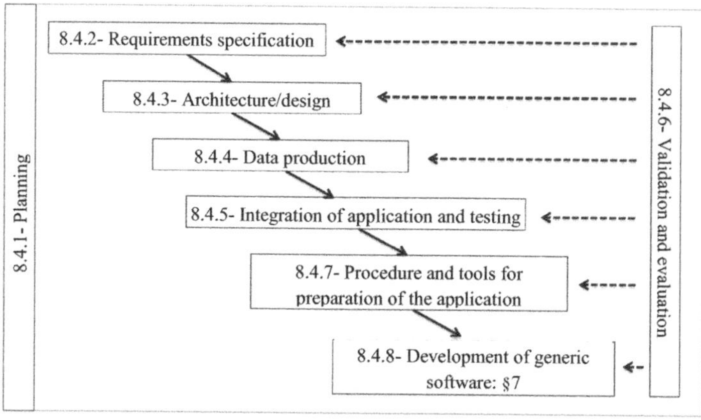

Figure 6.11. *Data preparation cycle*

The data preparation cycle is divided into eight phases. Of those eight, there are three phases of particular importance:

– the planning phase (step 8.4.1), which is aimed at defining the data preparation strategy, the activities, the documentation needing to be produced and the necessary resources;

– the validation and evaluation phase (step 8.4.6), which is not really a phase in the truest sense of the word. Indeed, the only article associated with it says that it is important to monitor the performance of each phase of the realization cycle. As the implementation of ISO 9001:2008 is requisite, this phase is covered by the activities of verification which need to be put in place. That is why, in Figure 6.11, a checking activity is shown to the right of the end of phase;

– the phase of development of the generic application (step 8.4.8), which invokes section 7 of the standard and introduces stipulations relating to the setting of the parameters.

6.6.2.2. *Problem of circularity*

In section 8.2, which relates to the input documents, it is indicated that, for the input to the data production process, it is necessary to have the requirements specification, the architecture, the application conditions and the user manual for the generic software, which implies that the generic software has already been realized and the data preparation process remains to be defined.

Section 8.8 of the standard pertains to the development of the generic software. In directly invokes section 7 after having identified the additional recommendations linked to the parameter data. As Figure 6.12 shows, therefore, there seems to be a circular aspect to the process.

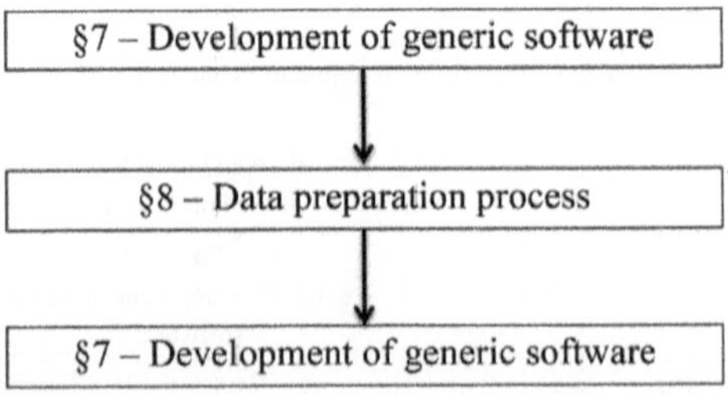

Figure 6.12. *Circularity between the generic software and the data preparation*

In fact, the circularity stems from the fact that there are a number of possibilities for developing a software application and the parameter data, as shown by Figure 6.13. In case 1, we begin by realizing the generic application and on that basis, we can then develop the data. On the contrary, case 2 involves carrying out the data preparation first and then developing the generic application afterwards.

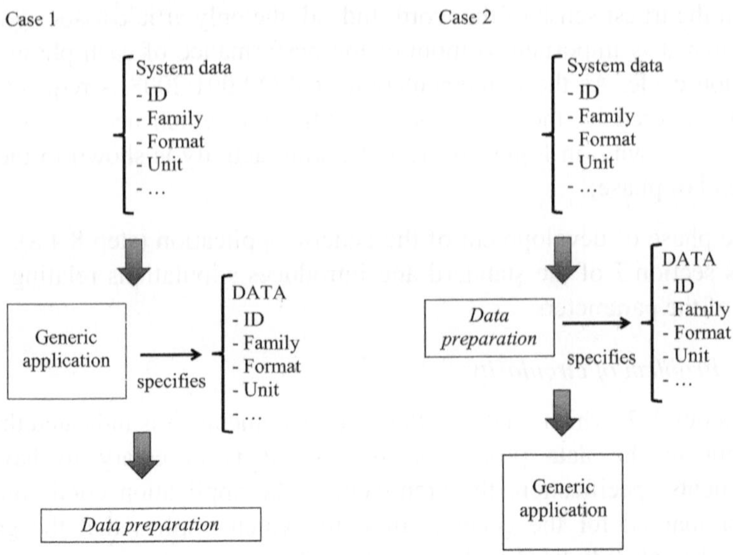

Figure 6.13. *Various realization processes*

From another standpoint, in the railway domain, at present, there are two approaches to the format of parameter data:

– the first approach, which corresponds to case 1 in Figure 6.13, is called "tabular data": on the basis of the algorithms identified in the generic software, we construct tables of specific data devoted to the algorithms. This approach helps to optimize the execution times and decrease the complexity of the algorithms, but it is difficult to find the system data in the tables, and the number of time data units then becomes extremely great. This type of approach has been used on the SACEM [GEO 90], MAGGALY, the SAET-METEOR system [BOU 00, BOU 06, DEL 99], the VAL at Charles de Gaulle Airport (see Introduction), etc.;

– the second approach, which corresponds to case 2 in Figure 6.13, is called "industrial data": on the basis of the input documents, we produce files of parameter data which are centered on the industry, containing the objects of the system. A high-level language (state machine diagrams, Petri nets, etc.) is used to describe each object, the different states and the links with the other objects. The generic software, therefore, is a machine for executing the objects, which must respect the semantics of the high-level language that has been used. This type of approaches has been employed for New-Generation Signaling Stations (known as PAINGs, from the French term) – [BOU 10a], Chapters 4 and 5 – and CMPs [GAL 08].

1)

Name	1	2	...
Position	10	20	...

2)

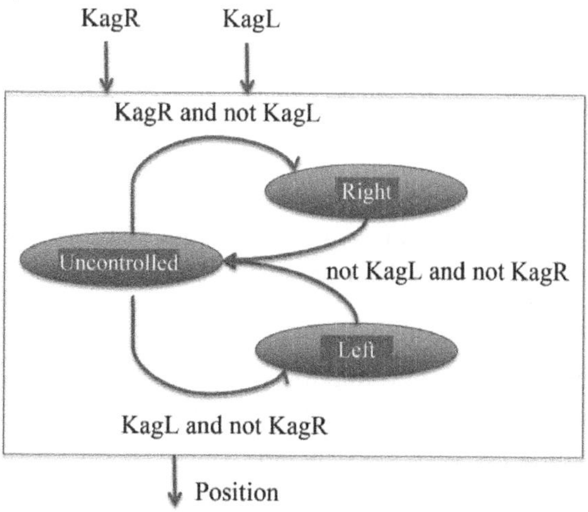

Figure 6.14. *Example of a parameter*

The choice of one of the two strategies determines the type of generic applications to be realized, and the overall process of realization of the final software application.

For Case 1 in Figure 6.13, the reality is a little more complex. In general, the data and the generic software are developed in parallel. This parallelism is caused by the fact that during the realization of the software, it may be necessary to introduce:

– specific data to control the algorithms (see, for example, the discussion in section 7.4.3);

– data with specific characteristics which help to optimize the software's execution and/or decrease the complexity of the algorithms. In the case of complex algorithms with multiple instances of element searches, it may be preferable to simplify the algorithms and generate tabulated data specific to that algorithm. In that case, the complexity is transferred from the algorithm executed by the generic software to the data production algorithm.

Finally, we can see that section 8.2 of CENELEC 50128:2011 needs to be supplemented by a set of system documents to identify the data. These system documents (System Data Specification, etc.) must describe the data from the point of view of the system (name, description, attributes, characteristics, unit, type, etc.).

6.7. Data preparation process

6.7.1. *Management of the data preparation process*

6.7.1.1. *Planning*

In this section, we shall discuss step 8.4.1, which is called "planning". It is necessary to draw up an application preparation plan (APP).

Figure 6.15 shows that data preparation involves managing two types of activities: the production of the parameter data and the integration of those data with the generic software. For this reason, the APP needs to be realized either for each specific application or for a category of specific application. If it is produced for a category of specific application, then we shall finally validate a family of configurations.

As with the software quality assurance plan (SQAP), the APP must:

– choose between industrial data and tabulated data;

– describe the process of development of the application (see sections 6.7.1.2 and 6.7.1.3);

– describe the human resources, the organization and the forms of independence;

– for each phase of the process, identify the activities to be performed, the input and output documents and the human resources needing to be mobilized;

– for each phase of the process, identify the verifications to be carried out;

– identify the documentary structure;

– identify the tools used;

– describe the principles of the configuration management.

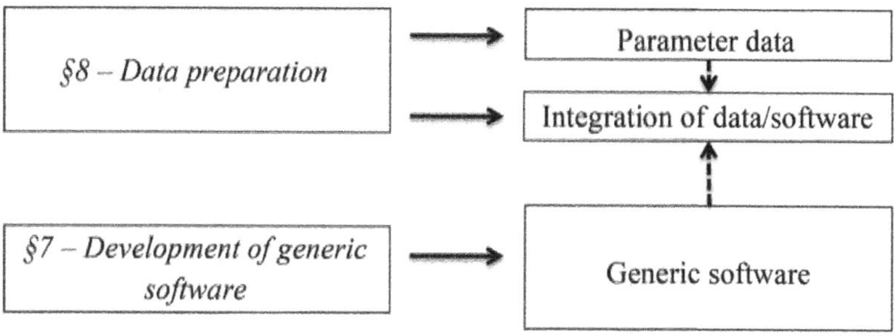

Figure 6.15. *Link between the two realization processes*

Software assurance, as defined in Clause 6 of CENELEC 50128:2011, identifies a tree diagram of plans (see Figure 4.5) which may or may not be shared with the data preparation. Thus, the APP may comprise a single document or be based on other documents (SQAP, SDP, SCMP, VVP, etc.).

Measure	SSIL0	SSIL1; SSIL2	SSIL3; SSIL4
1. Tabular specification methods	R	R	R
2. Application-specific language	R	R	R
3. Simulation	R	HR	HR
4. Functional testing	M	M	M
5. Checklists	R	HR	M
6. Fagan inspection	–	R	R
7. Formal design review	R	HR	HR
8. Formal proof of the correctness of the data	–	–	HR
9. Walkthrough	R	R	HR

Table 6.1. *Table A.11 from CENELEC 50128:2011*

CENELEC 50128:2011 identifies a number of techniques to be employed for data preparation, in its table A.11 (see Table 6.1 above). Table A.11 of the standard outlines two types of techniques:

– languages (1 and 2);

– means of verification (3 to 9)[4].

One of the problems of data preparation relates to the means of expression for specifying the data (Clause 8.4.1.11, Table A.11, items 1 and 2). Indeed, the data may be in different forms – either tabulated data or industrial data (state machines, Petri nets, etc.).

The APP therefore needs to identify the formalisms used to describe the data and the data production process, so that it is feasible, comprehensible, limits the introduction of errors and is reproducible and maintainable.

6.7.1.2. *Integration between the data and the generic software*

The link between the data and the generic software is an important point, which needs to be properly grasped. As the generic software consumes data, it can export constraints (see Clause 8.4.16 of the standard) which the data must respect (e.g. "always finish a table with the value –1", "express

4 See Chapter 4 for a description of the verification techniques.

distances in cm", etc.), and these constraints can be directly linked to the software safety.

There are different levels of integration between the parameter data and the generic software, as illustrated by Figure 6.16. Case 1 presents an example of software where the parameter data are dispersed or fully integrated into the software. In case 2, the parameter data are a contiguous whole, but are still integrated in the application. Then we have case 3, where the parameter data are outside of the software.

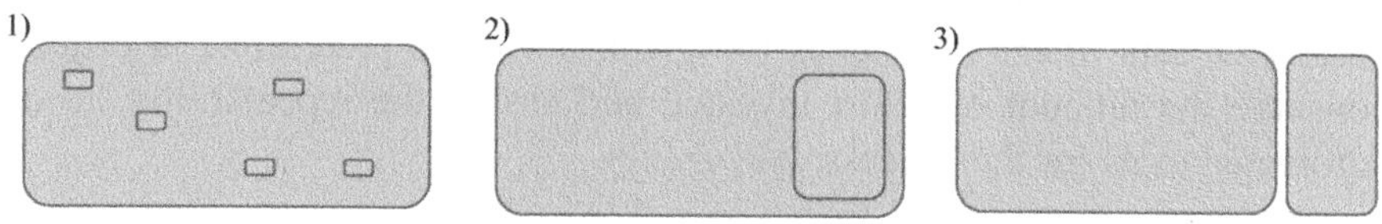

Figure 6.16. *Software integration levels*

The identification of the parameter data, the principle of separation between the data and the generic software and the interfaces must be identified at the point of specification of the generic software (Clauses 8.4.8.2 and 8.4.8.3 of the standard).

Case 3 is interesting (see Clause 8.4.8.4 of the standard) because it simplifies the maintenance activities – a modification of the data can no longer lead to recompilation – but it induces problems of controlling the integrity and validity of the data. Indeed, the protective measures in place for the executable no longer apply for the software, so we need to put specific protective measures in place for the parameter data, and mechanisms for the executable to check the validity of the data. The problem is: "how can the software know that an integral configuration (a coherent dataset) is an appropriate configuration for it?".

6.7.1.3. *Data production process*

The data production process is designed to transform (see Figure 6.17) system-level data into data that can be manipulated by the software. This transformation can be performed manually or it can be automated.

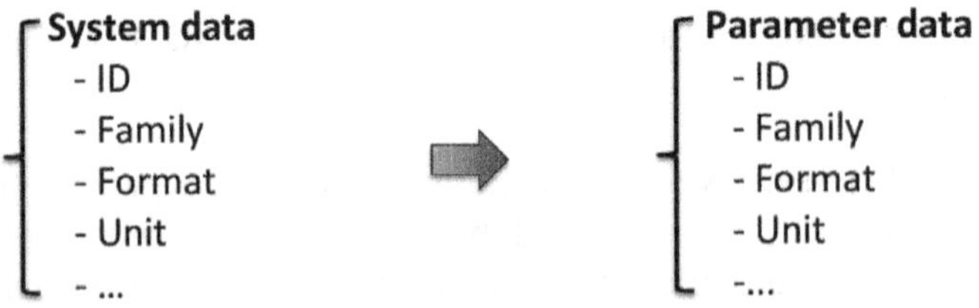

Figure 6.17. *Data transformation*

If the number of data units is small, it is possible to manually produce the parameter data. If the number is too great, it will be preferable, in order to conserve the amount of effort invested and ensure the repeatability, to use automated tools for the process.

The specification of the generic software must identify the SSIL, the software requirements and the data. Hence, it is possible to identify the SIL of the parameter data.

Now we need to define the data production process. The main difficulty relates to the correctness of the transformation; with this in mind, it is possible to use three approaches [BOU 07] for data production (see Figure 6.18).

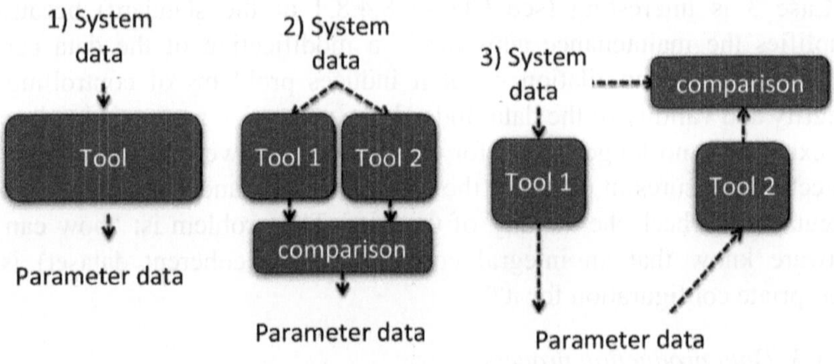

Figure 6.18. *Data transformation*

The first case consists of developing only one tool to produce the data. The second case is to use three tools: two diversified tools – but which have the same specification – perform the same transformation, and a comparison engine is used to check that the result is identical (or similar in the case of partial comparison). The last case is to realize a tool which performs the transformation and a tool which reverses the transformation, and finally use a comparison engine to check that the result is similar (generally speaking, transformations do not preserve all the information, so it is impossible to verify absolute equality).

The third case in Figure 6.18 ensures greater diversity (two different specifications) between the two tools than does the second case (same specification).

Thus, we have at least four possible approaches, but they must be combined with the possibility of manually checking the data, so we have Table 6.2.

Production	Verification of data	Characteristics	Comment
Manual	Manual (walkthrough, etc.)	Not much data	The number of data units is reasonable, and there is no need for automation
Manual	Tests	Not much data	It is possible to show that the data are correct by whole-program software tests This approach is not practicable in case of a large number of data units
Automated – case 1	Manual (walkthrough, etc.)	Not much data	The small number of data units means it is possible to carry out a manual check
Automated – case 1		Data SSIL2 or SSIL3-4	A unique tool developed for the data SSIL
Automated – case 2		Data SSIL2 or SSIL3-4	We have a double chain, which enables us to use tools from a lower SSIL (diversification function) than the SSIL expected for the data
Automated – case 3		Data SSIL2 or SSIL3-4	We have a highly-diversified double chain, which enables us to use tools from a lower SSIL (diversification function) than the SSIL expected for the data
–	Automated	Data SSIL2 or SSIL3-4	The safety of the data may be based on the implementation of a tool to verify the data (e.g. by proof)

Table 6.2. *Combinations to realize a data production process*

The APP must describe the process of data production needing to be implemented as a function of the SSIL, and identify the set of tools that must be employed. For each tool, either it needs to be created or it is re-used. In all cases (Clause 8.4.1.10 in the standard), SSIL objectives and a class of qualification (T1, T2, T3) must be associated with the tools, and a certification of the tools must be conducted in accordance with Clause 6.7 of CENELEC 50128:2011 (see Chapter 8).

The 2011 version of CENELEC EN 50128 formally introduces the need to qualify the tools (see section 6.7 of the standard and Chapter 9 of this book).

6.7.1.4. *Inputs to the data preparation process*

As explained in the previous sections, section 8.2 of CENELEC 50128:2011 does not identify all of the input documents for the data production process. To begin with, for the inputs, it is crucial to have the system documents mentioned below.

These documents must take account of the system data – including, for example:

– system requirements specification;

– description of system architecture;

– system safety requirements specification;

– etc.

In addition, depending on the case from Figure 6.18 that applies, it is essential to have at least a software quality assurance plan (SQAP), which describes the process of realization of the specific application, the steps (software development, development of the data preparation process, data production, etc.) and the order of those steps.

6.7.1.5. *Study of safety of data process*

The data production process is based on tools (see Figure 6.18), which are either reused from a previous project or need to be developed. The data are manipulated by a generic software application, developed with a particular SSIL; thus, the data must be equally as safe as the software by which they are consumed.

Clause 8.4.1.8 recommends that a safety analysis be performed of the proposed data preparation process, with a view to confirming that we can achieve the expected level of safety for the data. This safety analysis would help confirm the levels of qualification expected for the tools and the objective SSILs allocated to them.

6.7.1.6. *Overview*

Ultimately (see Figure 6.19), before being able to produce the APP, it is essential to have an SQAP which describes the overall strategy to realize the specific application (development of a generic application, development of the data, instantiation of a configuration, creation of the specific application) and the order in which these activities take place. This SQAP must also identify the nature of the data (tabulated or industrial) and the principles of integration of the data with the generic software.

On the basis of the SQAP, it is possible to produce the APP, which must – in addition to defining the organization, the process and the resources – identify the data preparation process (see Table 6.2), identify the tools which can be reused and those which need to be developed, and identifier the level of qualification of the tools and their objective SSILs (see the safety study).

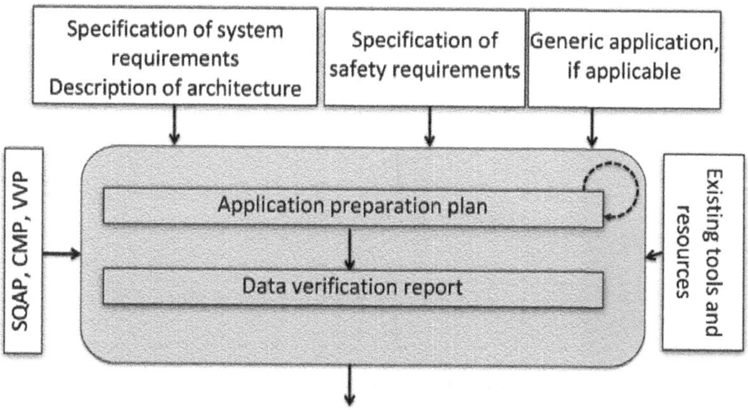

Figure 6.19. *APP and its verification*

As is stipulated by ISO 9001:2008, each product must be verified. The CENELEC 50128:2011 standard identifies a data verification report (DVR).

The APP must be verified (for conformity to CENELEC 50128:2011, coherence and implementation on each specific application). The results of this verification must be reported in the DVR.

6.7.2. *Verification*

In Chapter 4, we discussed software assurance. Amongst other things, software assurance includes V&V. Thus, section 4.7 is applicable both to the generic software and to parameter management. Therefore, it is not necessary to repeat the presentation of the activities of verification.

6.7.3. *Specification phase*

6.7.3.1. *Description of the step*

As Figure 6.20 shows, the input to the specification phase is the planning documents (APP and other plans), the system requirements, the system architecture and the safety requirements.

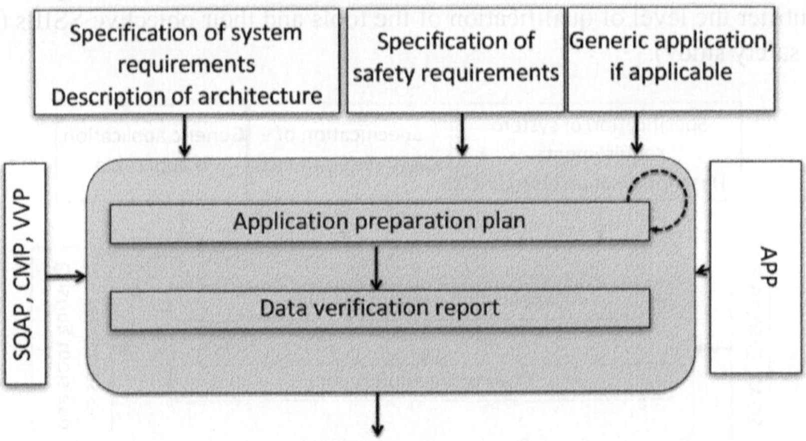

Figure 6.20. *Specification phase*

6.7.3.2. *Specification of application requirements*

The specification phase involves producing an application requirements specification (ARS), which must cover both the data and the generic software. For the generic software, if it is developed, we need to take account of the constraints which it exports. If not, we need to describe the requirements that characterize it.

The ARS must, firstly, contain a description of the external interfaces. Figure 6.21 illustrates the system input interfaces (all of the system data and their description), the auxiliary interfaces which are the existing tools, and the output interfaces, which are the data produced for the different subsystems and devices (see Figures 6.7 and 6.8). This view of the process enables us to take account of the issue of compatibility of the data produced for the different subsystems. If different data production processes are employed (one for each subsystem and/or device), it is crucial to put in place a process to ensure compatibility and coherence of the data.

This specification must take account of the choices that have been made in the APP with regard to the data production process (see Table 6.2 and the accompanying discussion). We must characterize the input data and the output data, clearly indicating the desired properties (functional and non-functional requirements) regarding those data. Note that the SSIL associated with the data must be identified.

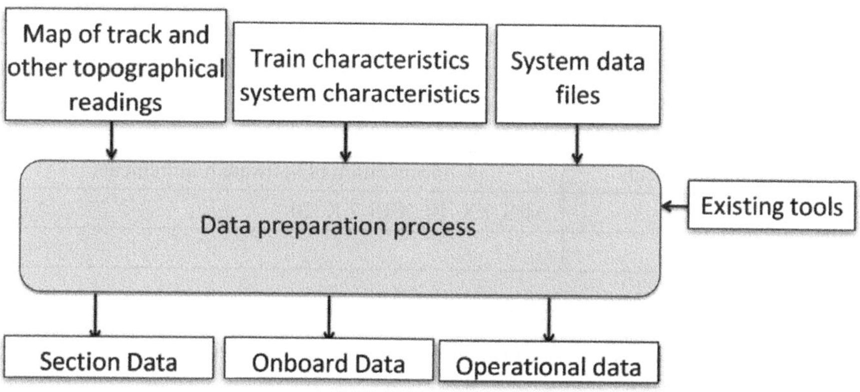

Figure 6.21. *Specification phase*

The ARS must take account of the separation between the generic software and the data, and therefore must identify the means of protection of the data and the principles of interfacing between the parameter data and the generic software.

For each requirement, we must identify:

– the link (traceability) with the phase input requirements;

– the safety attribute (yes or no);

– the verification attribute, which indicates whether the requirement is testable or whether a specific analysis is necessary.

With regard to the testability of the requirements, it must be remembered that all the requirements must be verifiable and that some of them must be testable. In addition, it is necessary to verify that the specification is:

– complete;

– coherent;

– comprehensive, clear and precise;

– verifiable;

– maintainable;

– traceable.

Table 6.3 shows an example of traceability between the input requirements and the software requirements. In the example, we can see a system requirement that is not fulfilled; it is therefore necessary to add a justification in order to prove that nothing has been overlooked.

System requirements	Specification of software requirements
SyRS_EX_1	ARS_EX_10, ARS_EX_20
SyRS_EX_2	–
	…

Table 6.3. *Example of traceability between the system- and software specifications*

Table 6.3 is not sufficient, because we need to identify the software requirements which are not traceable (in the aeronautic domain, the term used is "derived requirements"). Therefore, it is necessary to add inverse traceability.

6.7.3.3. *Complete data preparation tests*

The standard does not require us to prepare the complete application test specification (CATS). We recommend – in order to ensure that the test cases are pertinent – performing a specification of the complete application tests whose objective is to cover all the requirements in the ARS. As the only input element at this level is the ARS, we are indeed in the process of specifying black-box tests (i.e. tests performed with no knowledge of the realization).

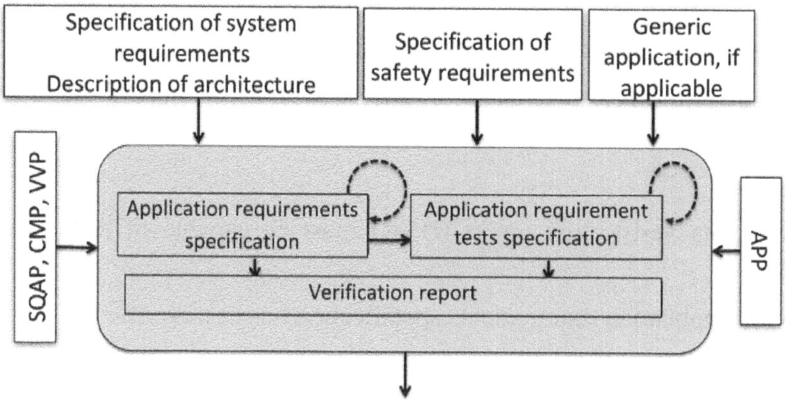

Figure 6.22. *Specification phase*

Figure 6.22 shows the step of specification of the application after addition of the CATS. The purpose of the CATS is to demonstrate that the process of parameter-setting carried out conforms to the stated requirements. The writing of the CATS enables us to show that the requirements are testable. Indeed, the person in charge of producing the CATS must perform a detailed requirements analysis (identification of the boundary values for each input, identification of equivalence classes and identification of observables).

The objective of the CATS is to cover 100% of the requirements. For each non-testable requirement, we must have a justification and an alternative verification activity. The CATS must contain a traceability matrix demonstrating that the requirements have properly been tested and/or verified, but we must also show that each test case is associated with a requirement.

The realization of the CATS, therefore, is a verification of the software requirements specification.

6.7.3.4. *Verification*

The verification of the ARS (and of the CATS) must relate to:

– the readability of the requirements;

– the traceability of the requirements;

– the internal coherence of the requirements;

– the respect of the project quality plans;

– the conformity to CENELEC 50128:2011, with account taken of the fact that the data preparation process will be based on existing tools or tools to be developed.

The data verification report (DVR) must be capable of demonstrating that:

– the application requirements specification has been verified;

– the whole-application test specification (if it is realized) has been verified;

– the processes defined in the APP and in the other plans (SQAP, VVP, etc.) have been applied properly.

The specification phase verification report can thus be formalized as the end-of-phase report, and be written following a formal review referencing all the records of the verification (list of input documents, completed checklists, review report, specific analysis, list of anomaly sheets, etc.). The aim of the formal review is to provide authorization to proceed to the next phase (with or without reservations).

6.7.4. *Architecture phase*

6.7.4.1. *Description of the architecture and the design*

Once the application requirements specification has been performed, it is then possible to put a form of architecture in place (see Figure 6.23), but in this case it is the architecture of the data production process.

On the basis of Table 6.2 and Figure 6.18, we must identify the tools that need to be implemented, any differences in the hardware (it may be necessary to execute several of the tools or a process) and the human resources. This architecture must take account of the results of the process safety study.

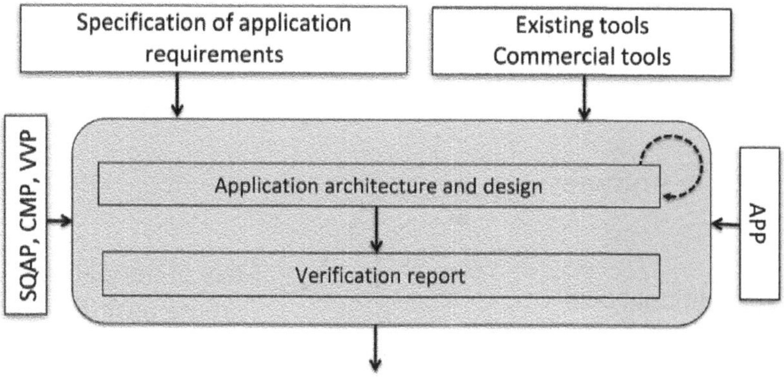

Figure 6.23. *Architecture phase*

On the basis of the list of tools to be implemented, we identify the tools needing to be developed, to commercial tools and the tools that can be reused. For each tool, we must identify the need for qualification (category Tx) and the objective SSIL. For the tools needing to be developed, the algorithms or transformations (see Figure 6.24) to be implemented must be specified.

To conclude, the application architecture and design (AAD) comprises:

– a description of the input data (files, etc.);

– a description of the intermediary data (files, formats, etc.);

– a model of the data preparation process;

– the hardware resources to be implemented;

– the human resources to be implemented;

– a list of tools with, for each tool:

  - the list of requirements,

  - the list of interfaces,

- the Tx category and the objective SSILs,
- the algorithms to be used;

– one or more traceability matrices.

One of the important subjects at this stage relates to the production of the intermediary files and their management (which configuration management, what type of protection, etc.).

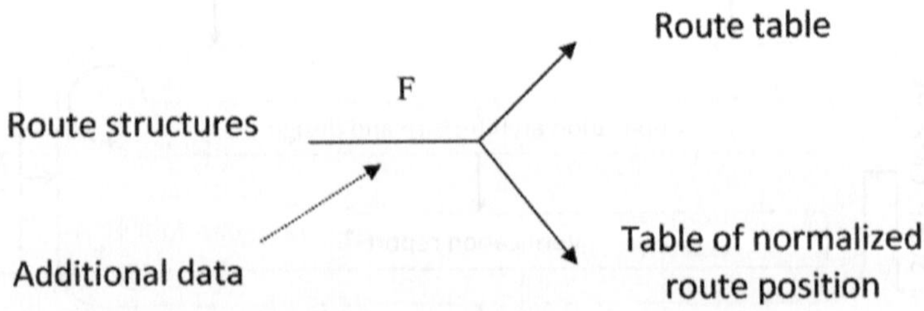

Figure 6.24. *Example of transformation*

6.7.4.2. *Integration tests*

The new version of CENELEC 50128 does not stipulate the preparation of any integration tests. In view of the complexity of the data preparation process (multiple tools, different machines and various people involved), it is preferable to conduct software/software and software/hardware integration tests. The complete architecture phase is shown in Figure 6.25.

The S/S[5] integration tests have the aim of verifying that the exchanges between the tools function correctly (the correct filenames, the right format, true accuracy of the data, effective protection of the files, etc.). The objective of the H/S integration tests is to demonstrate that the software applications function correctly (proper operation, no overconsumption of memory, acceptable memory consumption and execution times, etc.).

It is possible to create two different documents or a single document. We speak of the AITS (Application Integration Test Specification). The aim of the AITS is to demonstrate that the software/software and software/hardware interfaces are correct.

5 S/S for Software/Software and H/S for Hardware/Software.

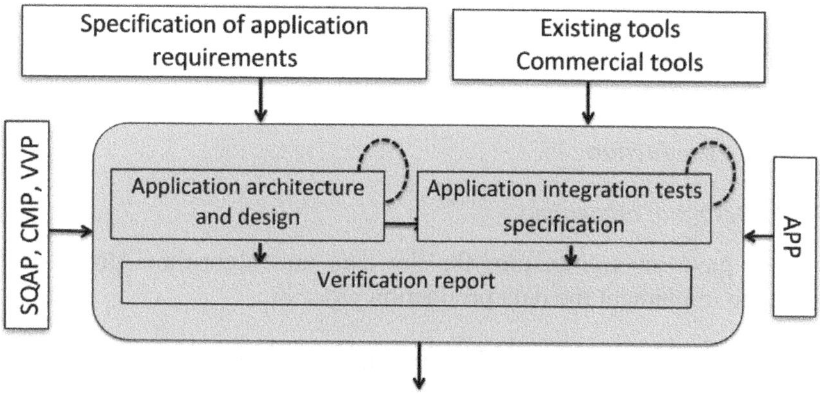

Figure 6.25. *Architecture phase with integration tests*

As the only input elements, at this point in the project, are descriptive documents, we are specifying black-box (with no knowledge of the realization).

The realization of the AITS is therefore a verification of the AAD.

6.7.4.3. *Verification report*

As previously indicated, CENELEC 50128:2011 identifies a verification report for the end of each phase. That verification report must enable us to demonstrate that:

– the specification of the application architecture has been fulfilled;

– the specification of the application integration tests (if done) has been fulfilled;

– the processes defined in the plans (SQAP, VVP, etc.) have been appropriately applied.

The verification report for the architecture and design phases can thus be formalized as the report on the end-of-phase review, and be written following a formal review, referencing all records of the verification (list of input documents, completed checklists, review report, specific analysis, list of anomaly sheets, etc.). The aim of the formal review is to provide authorization to proceed to the next phase (with or without reservations).

6.7.5. *Data production*

6.7.5.1. *Design of algorithms*

If we have an architecture for the data and algorithms process, it is possible to implement the data production process.

Figure 6.26 recaps that the data production process must take account of two aspects:

– the data production;

– the production of the final executable (data + generic software).

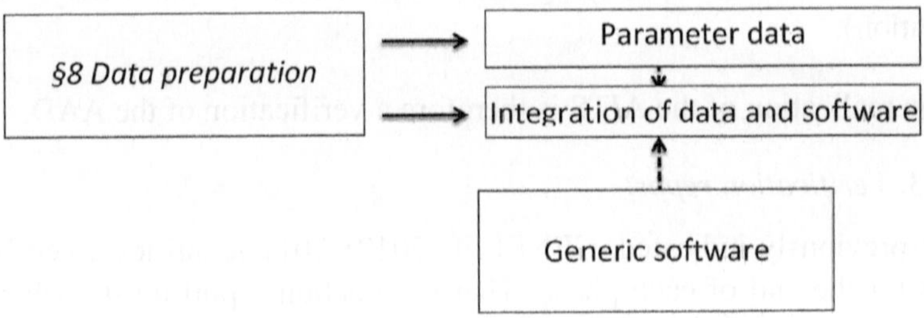

Figure 6.26. *Link between the two realization processes*

As regards data production, Clause 8.4.4.1 indicates that agrammatic languages are recommended for producing the data. There is a link in Table A.16 of CENELEC 50128:2011 (see Table 6.4). This recommendation supplements rows 1 and 2 in Table A.11, which spoke of modeling based on tables and specific languages.

As regards Table 6.4, rows 1, 2 and 3 make a direct link to section 3 of IEC 61131:2003[6] [IEC 03]. Row 4 talks about state graphs, but should be

6 IEC 61131 [IEC 03] introduces syntax and the semantics of programming languages for PLCs.

read as "state/transition graphs" in a broader sense (see Figure 6.14). Remember that this version of the standard indicates that it is possible to use other languages not mentioned here, after demonstration of their adequacy for the application.

Finally, the standard recommends using a graphic language close to those used in the particular domain (preserving the specific points of signaling, for instance, in order to facilitate verification by the experts in the domain) to describe the data and describe the transformations necessary to produce the application data.

Measure	SSIL0	SSIL1; SSIL2	SSIL3; SSIL4
1. Functional block diagram	R	R	R
2. Sequential function charts	–	HR	HR
3. Ladder diagrams	R	R	R
4. State charts	R	HR	HR

Table 6.4. *Table A.16 from the standard EN 50128:2011*

At this level, CENELEC 50128:2011 is not overly precise as to the documentation needing to be produced. Existing tools (such COTS tools or tools reused from former projects) can be employed, some tools may have been realized (see Clause 8.4.7) and a specific form of data coding may have been implemented. If we wish to be able to specify and perform tests (see Clauses 8.4.4.2, 8.4.4.3, etc.), it is necessary to have design documents which enable us to ensure the software maintenance. Thus, we feel it necessary to have at least one design document. Therefore, we recommend producing detailed design documents for each tool or algorithm being coded.

Figure 6.27 shows that on the basis of the document describing the architecture, it is possible to identify the tools which can be reused (COTS or existing tools) and that the qualification case (QC) must be able to demonstrate the appropriateness of the tool for the needs. If a development is necessary, a detailed design (DD) document and associated tests (ATS) (see the next section) need producing.

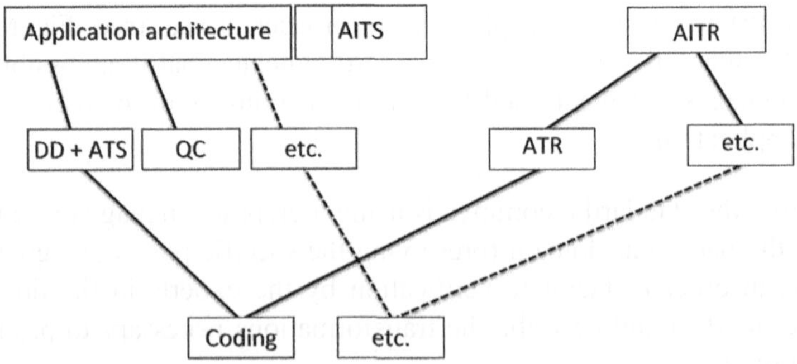

Figure 6.27. *Realization of data production process*

In case of tool development, the DD should contain the necessary requirement (functional and non-functional) to permit the development of the tools.

6.7.5.2. *Algorithm tests*

The 2011 version of the standard offers an improvement to deal with data production, but points remain to be clarified – e.g. the algorithm tests and design tests. Clauses 8.4.4.2 and 8.4.4.3 discuss the application test report, whereas the application test specification (ATS) is only defined later on (Clauses 8.4.4.5 and 8.4.4.6).

The ATS must enable us to show that the data and/or the algorithms do conform to the requirements and the architecture. The best option, therefore, is to have a design document which identifies the needs and can serve for the coding and test production.

In the previous sections, we introduced overall tests and integration tests. At this level, it would be preferable to introduce tests of the algorithms and/or data designed to show that the functionalities are present in the tools used and/or that the algorithms developed function properly. Thus, the ATS would more likely pertain to the type of component tests or module tests. The realization of the ATS, therefore, is a verification of the design documents (if they exist).

Provided the algorithms or data are testable, we must execute the tests described in the ATS and write an application test report (ATR) that documents the test results (version, date, tester's name, etc.).

To conclude this section, we can state that the ensemble comprising the CATS, the AITS and the ATS enables us to cater for Clause 8.4.4.5 and the ensemble comprising the CATR, the AITR and the ATR enables us to cater for Clause 8.4.4.2.

6.7.5.3. *Data verification*

Clause 8.4.4.4 introduces a specific application preparation verification report (APVR). The verification of the algorithms and/or data is done by way of a walkthrough, which involves verifying (see Figure 6.27):

– the existence of the data handled by the elaboration process;

– the existence of the data handled by the software;

– the existence and correctness of the generation process;

– the traceability of the requirements that apply to the data.

Activities of verification of the algorithms and/or data should be performed in order to demonstrate that the data are correct and complete. These activities must be documented in the APVR. These verifications may be introduced into the process (double execution, specific verification tool, property demonstration tool, etc.) or be carried out in the form of specific checks or tests.

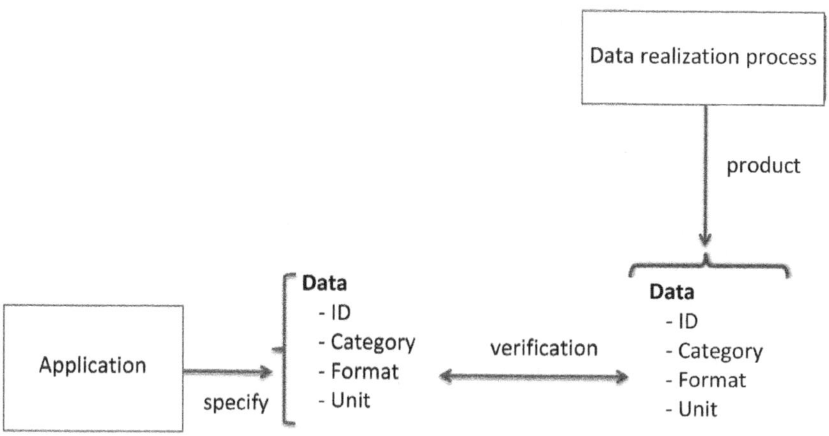

Figure 6.28. *Data compatibilities*

6.7.5.4. *Activity of verification*

As indicated above, the activity needs to be verified, so a specific VR needs to be produced or the overall VR needs to be updated (if all the verifications are in the same report), and checks must be performed in order to demonstrate that:

– the processes defined in the APP and in the other plans (SQAP, VVP, etc.) have been properly applied;

– the design of the algorithms and/or data has been verified;

– the specification of the algorithm tests (if done) has been fulfilled (e.g. with a requirement fulfillment analysis);

– the algorithm test report (if done) has been fulfilled;

– the verification of the data (APVR) has been performed.

The algorithm and/or data design phase report can thus be formalized as the end-of-phase report, and be written following a formal review referencing all the records of the verification (list of input documents, completed checklists, review report, specific analysis, list of anomaly sheets, etc.). The aim of the formal review is to provide authorization to proceed to the next phase (with or without reservations).

6.7.5.5. *Overview*

The algorithm and data design phase is fairly complex, as is illustrated by Figure 6.29. The CENELEC 50128:2011 standard is, indeed, rather ambiguous as to the activities that need to be established.

Figure 6.29 illustrates only the application tests (linked to the algorithms and/or data), but if we have put integration tests (AITS) and whole-application tests (CATS) in place, it is during this phase that we must (see Figure 6.27) integrate the tools to realize the data production process, and it is at this time that we must demonstrate that the process respects the requirements in its specification. The AITR and ATR therefore need to be produced and verified during this phase.

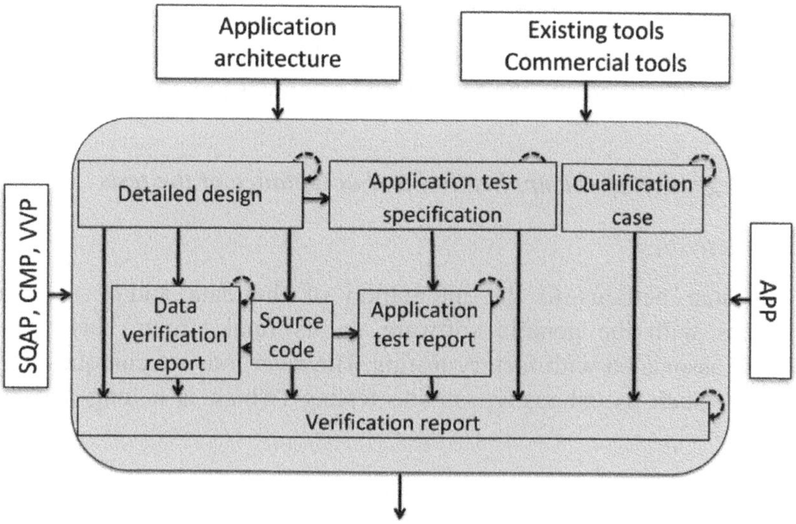

Figure 6.29. *Data and algorithm design phase*

In the context of a certifiable application, it is necessary to be able to repeat the tests in an identical manner. Therefore, the test realization process must be documented, the configuration of the input data and of the results must be managed, as must all elements necessary for the implementation of the tests.

The structure of the test reports (ATR, AITR and CATR) obeys the following plan:

– identification of the input elements;

– identification of the applicable plans (SQAP, etc.);

– resources used: people, tools, environment, etc.;

– version analyzed;

– list of test cases;

– demonstration of the achievement of the objectives;

– overview of the problems identified;

– conclusion as to the correctness of the version.

The overview of the problems identified requires that, for each failure that is detected, an anomaly file be opened. This anomaly file must describe the problem identified and the activity and resources mobilized to remedy it.

6.7.6. *Integration of the application and acceptance of the tests*

6.7.6.1. *Activities*

This step pertains to the integration of the data and/or application algorithms with the generic software on its final target. This phase is generally associated with factory testing. However, other techniques may be employed, such as exhaustive on-site testing (which is a long and costly process).

Clause 8.4.5.2 identifies the establishment of a specification for the application tests, which must demonstrate:

– the proper integration of the data and/or application algorithms with the generic software on its final target;

– the proper integration of the data and/or application algorithms with the complete installation.

It is regrettable that the standard identifies this document as being the same as that which was identified during the data production phase (previous section of this chapter). We recommend defining a specific test during called the installation test specification (ITS). One of the important inputs for the ITS is linked to the constraints exported by the generic software and by the data process.

The installation test report (ITR) needs to be drawn up at the end of the execution of the tests. It must clearly indicate what has been tested, by whom, when, and the results obtained.

6.7.6.2. *Verification of the activity*

As previously indicated, the activity must be verified, so a DVR must be produced, or the global DVR must be updated, and the checks should enable us to demonstrate that:

– the processes defined in the APP and in the other plans (SQAP, VVP, etc.) have been respected;

– the specification of the installation tests has been verified (e.g. with an analysis of the fulfillment of the requirements);

– the installation test report has been verified.

6.7.6.3. *Overview*

Figure 6.30 presents the complete step as stipulated.

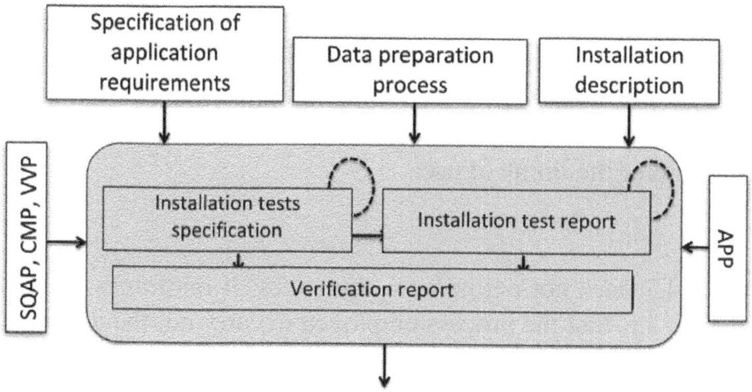

Figure 6.30. *Data and algorithm design phase*

6.7.7. *Validation and evaluation of the application*

Clause 8.4.6 recommends the establishment of verification activities for each phase. We have integrated that recommendation into each phase that we have presented. There is nothing more that needs to be realized.

6.7.8. *Procedure and tools for preparation of the application*

6.7.8.1. *Boundary*

Clause 8.4.7 of the standard covers a variety of subjects. It indicates:

– that the development of the specific tools must respect CENELEC 50128:2011;

– the necessity to validate and evaluate the compilation process, taking account of the data and/or the algorithms (a dedicated compilation chain may be necessary);

– that Clauses 9.1 (relating to deployment) and 9.2 (pertaining to maintenance) must be taken into account;

– that software assurance – as discussed in Chapter 5 – must be taken into account;

– that the application verification report must demonstrate the satisfaction of all the exported constraints, both by the generic software and by the data.

The finalization of the data production process and of the activities associated therewith needs to be formalized by the production of a software version sheet (SwVS). That SwVS must identify the configuration of the tools used, the documentary configuration and must contain the exported constraints and/or the limits of use.

6.7.8.2. *Overview*

Clause 8.4.7 does not define new objectives. It reinforces what was said in Chapter 4 – i.e. that the process employed to carry out the data preparation process must conform to ISO 9001:2008 and to software assurance as defined in Clause 6 of that standard.

The maintenance and deployment of the software are presented in Chapter 10.

6.7.9. *Development of generic software*

Clause 8.4.8 forms the link with Clause 7, which governs the development of the generic software. We presented the needs of that clause in the definition of the process of realization and writing of the APP, in section 4.3.1.

There are three important points:

– it is important to demonstrate that the whole-program tests on the generic software do indeed cover all the pertinent configurations of the data. If it is the case that not all the combinations have been tested, usage limits must be identified;

– as part of the software maintenance, we must demonstrate that all the modifications that have an impact on the generic software (or the configuration data) have been subject to an impact analysis in order to verify whether or not the configuration data (or the generic software) are impacted;

– it is necessary to demonstrate that the generic software and the configuration data are compatible.

The realization of a generic software application also generates a software version sheet. Previously, we identified the existence of the SwVS for the data production process. For a given installation, therefore, it is necessary to produce an SwVS which describes:

– the SwVS for the generic software used;

– the SwVS of the data production process;

– the configuration of the input data used.

6.8. Conclusion

CENELEC 50128:2011 deals with two aspects: the development of a generic software application (see section 7 therein) and the development of a process by which to set the parameters of a generic application (section 8).

In Chapter 7, we shall present the process of development of a generic software application. This chapter describes the process of data production. The data-production process is based on management of the software's safety (section 6 in CENELEC 50128:2011) and on the development and/or re-use of tools. The development of a tool can either obey the process laid out by CENELEC 50128:2011 or may be associated with a qualification (see Chapter 8).

We have presented the principles of data preparation, the constraints and our recommendations. Indeed, although section 8 of CENELEC 50128:2011 does represent real progress, it is still not sufficient to carry out a process. This is due to the complexity of the parameter-setting processes and to the vast number of possibilities.

6.9. Appendix: documentation to be produced

In this chapter, we have presented the realization of a parameter-based application. Here, we devote a small section to recap the list of quality plans which need to be produced as part of a project conducted in accordance with CENELEC 50128:2011.

Title	Acronym
Application Preparation Plan	APP
Application Requirements Specification	ARS
Complete Application Tests Specification	CATS
Application Tests Specification	ATS
Complete Application Tests Report	CATR
Application Architecture and Design	AAD
Application Integration Test Specification	AITS
Application Integration Test Report	AITR
Application Preparation Verification Report	APVR
Application Test Report	ATR
Source codes for the application data/algorithms	–
Application data/algorithm Verification Report	AVR
Installation Test specification	ITS
Installation Test Report	ITR

Table 6.5. *List of documents to be produced*

It should be noted that the list given in Table 6.5 is more complete than the standard, because the standard implicitly introduced some integration tests.

7

Generic Application

7.1. Introduction

As indicated in the previous chapters of this book, CENELEC 50128:2011 introduces the concept of generic software. In the final analysis, all software is considered to be generic, but certain programs have specifically set parameters. Chapter 7 of CENELEC 50128:2011 is given over to the realization of generic software.

The realization of a generic application can be done on the basis of the system elements (specification, architecture, knowledge of the hardware architecture, etc.) and/or after having identified the parameter data and the links with the software (see Chapter 6 of this book).

7.2. Software application realization process

This section is based on section 5.3.3 of the CENELEC standard. Figure 7.1 shows the V-lifecycle model as it is generally presented. The objective of the needs analysis is to verify the fulfillment of the client's expectations and the technological feasibility. The purpose of the specification phase is to describe what the software is meant to do (rather than exactly how it will do it). In the context of definition of the architecture, we seek to create a hierarchical breakdown of the software application into modules and/or components, and we identify the interfaces between those elements.

The description of each module/component (data, algorithms, etc.) is set out as part of the design. Often, the design phase is separated into two steps. The first, known as "preliminary design", involves identifying the data to be handled by the software and the necessary services; the second step – "detailed design" – involves describing all of the application's services by means of their algorithms. The design phase then leads into the coding phase.

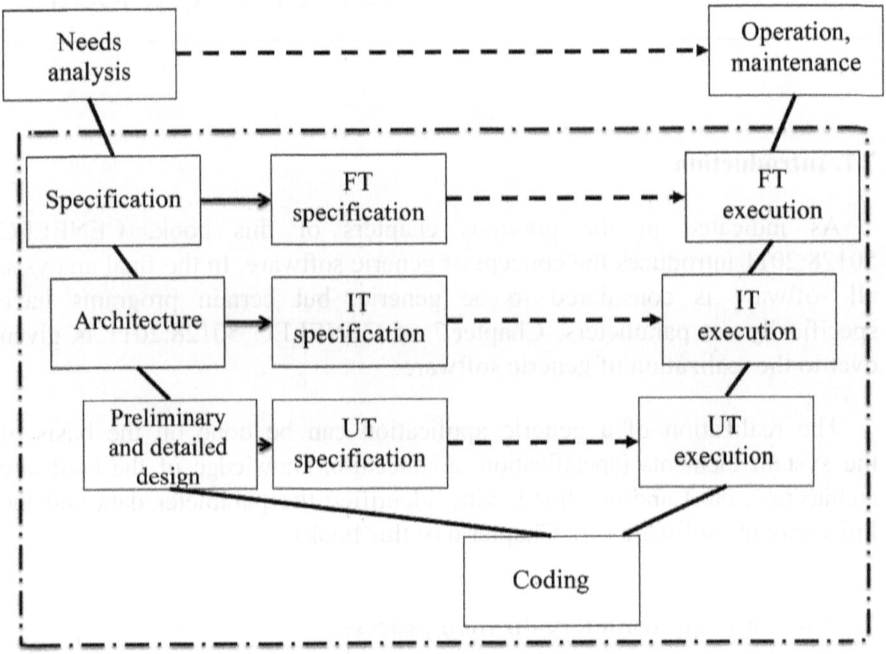

Figure 7.1. *V lifecycle including test specification*

Figure 7.1 shows that there are different phases of tests: unit tests (UTs in the figure), which focus on the lowest-level components; integration tests (ITs), which pertain to the software and/or hardware interfaces), and the functional tests (FTs), which aim to show that the product conforms to the set specifications. This approach of testing is compliant with the module decomposition. In the current approach, the unit is now the component and we do some component tests (CTs). As discussed in section 4.3.7, validation consists of demonstrating that we have created the proper product. It is for this reason that we speak not of validation tests (VTs) but of functional tests (FTs). In the new version of the standard, these tests are called overall software tests (OSTs).

As regards the operation/maintenance, it relates to the system's operational life and the handling of any future evolutions.

It should be noted that there is a horizontal correspondence (illustrated by the dotted arrows) between the activities of specification, design and the test activities. The V lifecycle is therefore divided into two phases: the "descendant" phase and the "ascendant" phase. The activities of the descendant phase must be prepared during the ascendant phase. Thus, Figure 7.1 is closer to the V lifecycle recommended.

Note that the V lifecycle illustrated by Figure 7.1 takes account of good practices. Initially, in the descendant phase, there was no production of specifications for the tests (for instance, see ISO 26262 [ISO 11], which stipulates that the test specifications be produced during the ascendant phase).

In the next few sections, we shall describe the different phases (specification, architecture, detailed design, coding, etc.), the methods, means and productions.

7.3. Realization of a generic application

7.3.1. *Specification phase*

7.3.1.1. *Description of the activity*

As Figure 7.2 illustrates, the input for the specification phase includes the planning documents, the system requirements, the system architecture and the safety requirements. On that basis, the first activity consists of setting out the specification for the software requirements, and the second of setting out the specification of the overall software tests.

Figure 7.2 identifies the verification phases (dotted arrows). Thus, we can see that the production of the overall software tests specification is an entirely separate verification. This verification is very important because if we can produce the OSTs, the requirements are feasible.

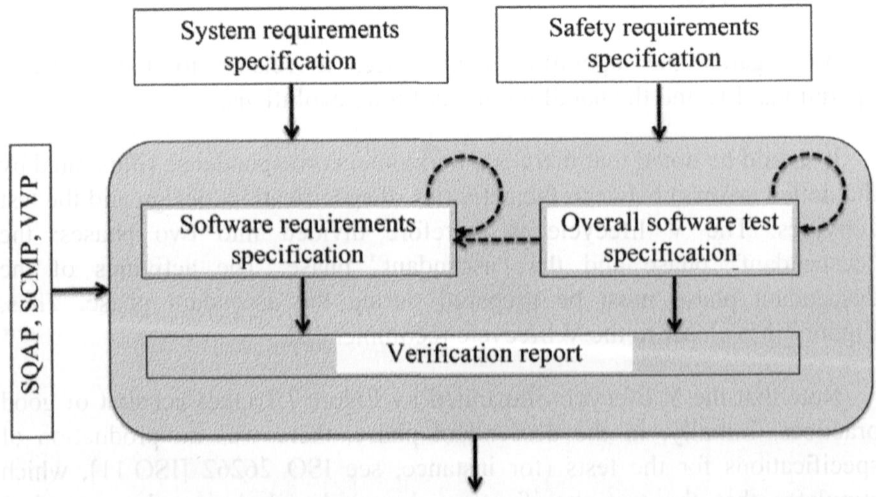

Figure 7.2. *The specification phase*

7.3.1.2. *Specification of software requirements*

7.3.1.2.1. Principles

The software requirements specification (SwRS) must, firstly, describe the boundary of the software. It is therefore necessary, on the basis of the elements of the system architecture and/or equipment, to identify the interfaces with the hardware and the software.

An interface diagram, similar to that shown in Figure 7.3, can be used to identify the external interfaces. We have two kinds of external interfaces: interfaces with other software (I_i, O_i) and interfaces with the hardware (Int_i).

The environment of the software application comprises interfaces with the hardware resources (memory, specific address, input/output, watchdog, etc.), with other software applications (baseware, connected application, etc.) and/or with the operating system.

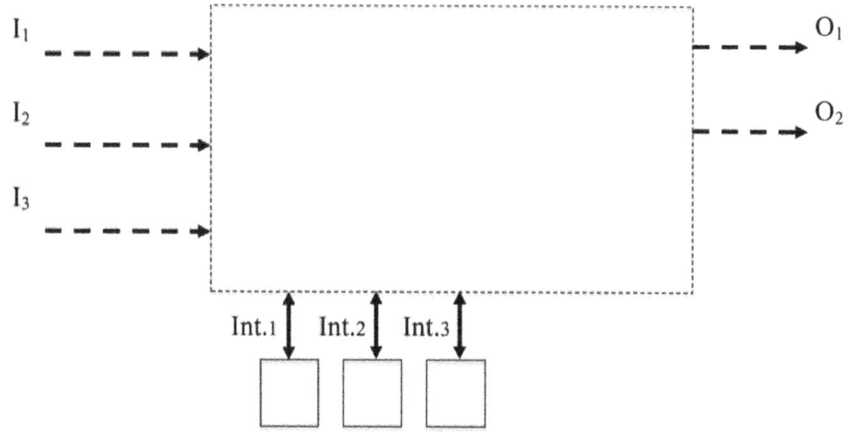

Figure 7.3. *Environment of a software application*[1]

For each external interface, we need to identify:

– name;

– objective;

– description;

– type: hardware, software;

– format, unit, precision;

– protocol if needed;

– data description: equivalence class, boundary value, behavior at boundary values and outside;

– memory management: buffer, stack with fixed size or not;

– time constraints;

– etc.

After description of the interfaces, it is necessary to identify the modes of operation (initialization, normal mode, defect mode, maintenance mode, etc.) and the transitions between the modes (see Figure 7.4).

1 I for Input, O for Output and Int for Interface.

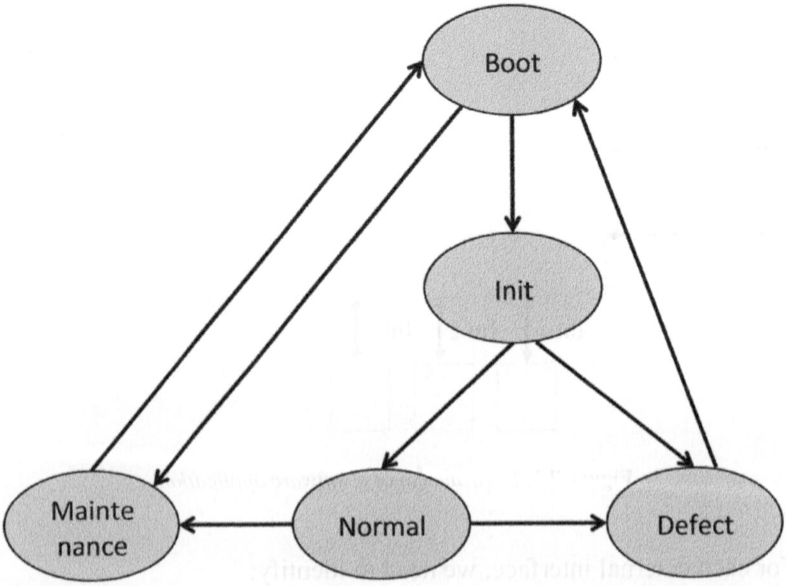

Figure 7.4. *Software operational mode*

For each operational mode, we need to identify the behavior and/or the list of function authorized. For example, in the maintenance mode, the main software cannot be executed and we can upload a new version of the software and download or erase the bug memory.

The software requirements specification must also identify the SSIL of each function. For this purpose, there are a variety of practices: either we indicate the SSIL of each function, or we indicate that the software is SSILx, or the SQAP of the project indicates the SSIL applied to the overall software.

On the basis of the interfaces and the modes of operation, it is possible to identify the software requirements. The software requirements must cover:

– the functional needs;

– the non-functional needs: safety, maintainability, etc.;

– the process and standard to be satisfied;

For each requirement, we must identify (see Table 7.1):

– the link (traceability) with the phase input requirements;

– the safety attribute (yes or no);

– the verification attribute, which indicates whether the requirement is testable or whether a specific verification analysis needs to be performed.

Attribute	Description
ID	Identifier
TEXT	Description of the requirement
SOURCE	Link to the upstream requirement
VERIFICATION	Activity to check the requirement : test or specific analysis
SAFETY	Yes or not
VERSION	Version associated with the requirement

Table 7.1. *Example of requirement attributes*

As regards the testability of the requirements, it must be remembered that all the requirements must be verifiable and that some of them are testable.

With regard to the requirements, we must verify that the specification is:

– complete;

– coherent;

– comprehensive, clear and accurate;

– verifiable;

– maintainable;

– feasible;

– traceable.

Table 7.2 shows an example of traceability between the input requirements and the software requirements. In the example, we can see a system requirement that is not covered; hence, it is necessary to add a justification to show that nothing has been overlooked.

System requirements	Specification of software requirements
SyRS_RQ_1	SwRS_RQ_10, SwRS_RQ_20,
SyRS_RQ_2	–
...	

Table 7.2. *Example of traceability for the requirements*

Table 7.2 is insufficient, because we need to identify the software requirements that are not traceable (in the aeronautic domain, we speak of derived requirements), so it is necessary to add inverse traceability.

As regards the completeness of the specification, it is not possible to obtain this property by analysis of the document. For the sake of the specification's coherence, it would be preferable to compare the requirements two by two, which is not always possible. In light of these two difficulties, it is recommended to create a model. The realization of a model provides a basis for the completeness and coherence analysis.

Technique/Measure	SSIL0	SSIL1; SSIL2	SSIL3; SSIL4
Formal methods (based on a mathematical approach)	–	R	HR
Modeling	R	R	HR
Structured methodology	R	R	HR
Decision table	R	R	HR

Table 7.3. *CENELEC 50128:2011 – Table A.2*

In the lower part of Table A.2 (see Table 7.3) of CENELEC 50128:2011, it is indicated that it is mandatory to have a textual expression of the requirements and that it is possible to associate with that text all the necessary modeling elements. As is shown by Figure 7.5, all or some of the requirements can be modeled in order to perform verifications of completeness and coherence.

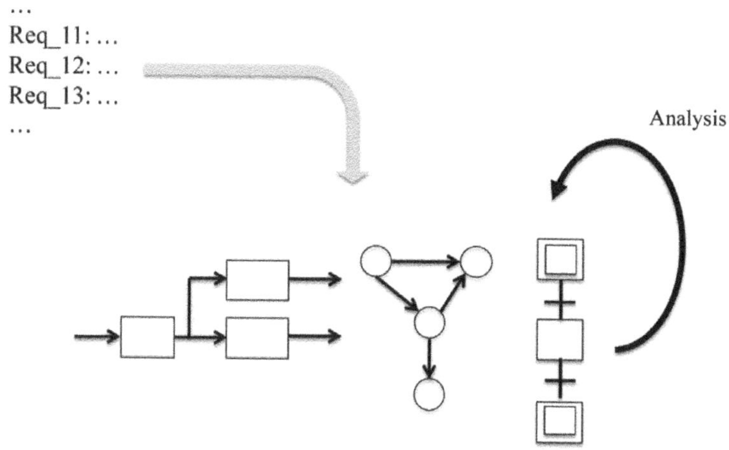

Figure 7.5. *Requirements, model and verification*

In conclusion, the software requirements specification is made up of:

– a description of the boundary of the system (hardware interface, software interface);

– a description of operational mode with an identification of the associated services;

– a list of the software requirements;

– one or more traceability matrices;

– one or more models.

7.3.1.2.2. Use of formal methods

From the old version of CENELEC 50128, formal methods were introduced as a necessity. One of the difficulties lies in the fact of having revealed the formal methods from the moment of the software requirements specification.

For our purposes, it is not possible to replace the specification of the requirements in natural language with a formal model. The formal model must remain an additional element to verify the requirements.

In Chapter 8, we shall give a more detailed presentation of the concepts of models, formal methods and formal techniques.

7.3.1.3. *Specification of overall software tests*

On the basis of the list of interfaces and the list of requirements from the software requirements specification, it is possible to prepare an overall software test specification (OSTS). As the only input element at this level is the software requirements specification (SwRS), we are specifying black-box tests (with no knowledge of the realization).

The aim of the OSTS is to show that the software being created does conform to the needs. However, the writing of the OSTS enables us to show that the requirements are testable and feasible. Indeed, the person in charge of producing the OSTS must perform a detailed analysis (see section 4.7.2.4) of the requirements (identification of the limit values for each input, identification of the equivalence classes and identification of the observables).

As Figure 7.6 shows, on the basis of the requirements, we identify test cases (TC_x). These test cases describe a situation to be attained, with that situation being linked to an equivalence class. Based on the test cases, it is possible to prepare the test scenarios. A test scenario describes a situation. Hence, a test case can be involved in a number of scenarios.

The preparation of the scenarios needs to be documented in the OSTS, and there is no need to have a computer file on the subject at this point in the project.

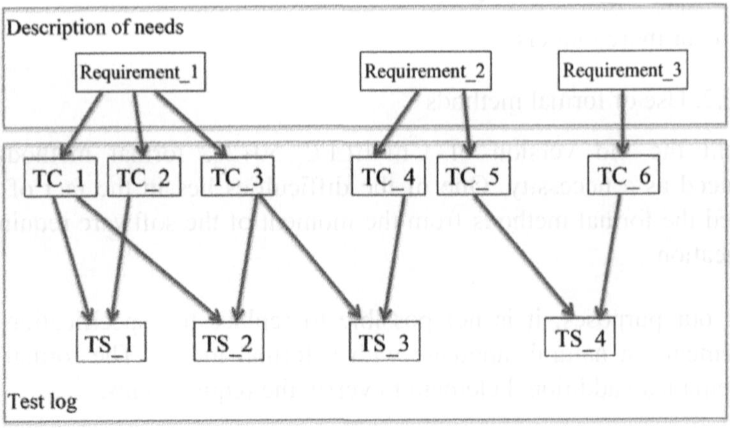

Figure 7.6. *Link between the requirements, the test cases and the test scenarios*

The aim of the OSTS is to cover 100% of the requirements. For each non-testable requirement, we must have a justification and an alternative verification activity. The OSTS must contain a traceability matrix demonstrating that the requirements have been tested and/or verified, but we must also demonstrate that each test case is associated with a requirement.

Thus, the realization of the OSTS is a verification of the software requirements specification.

7.3.1.4. *Verification report*

As required by ISO 9001:2008, each product must be verified. CENELEC 50128:2011 identifies an end-of-specification-phase report. This verification report must be able to demonstrate that:

– the software requirements specification has been satisfied;

– the specification of the overall software tests has been satisfied;

– the processes defined in the plans (SQAP, VVP, etc.) have been properly applied.

The verification of the SwRS and the OSTS must be carried out by way of a documentary review at least, in the knowledge that the realization of the OSTS is an entirely separate check. The review of the SwRS and OSTS must be performed with a specific objective in mind. That objective will need to be formalized by way of checklists.

On the output of the software requirement phase, we have three kinds of verification:

1) quality check: we need to verify that the procedure, plan and process are apply correctly;

2) technical check: we need to verify the technical content of the requirement;

3) safety check: we need to verify that the safety requirements in input are correctly assume

For the technical check of the SwRS, we need to verify:

– the correctness of each requirement;

– the coherency of each requirement (the requirement is correct);

– the completeness of each requirement (all items used are defined);

– all requirement are stand alone (not necessary to read many requirement to understand one);

– the requirement size;

– the requirement vocabulary

– the requirement specification completeness;

– the requirement specification correctness;

– the requirement coverage;

– if the requirement is verifiable.

It is possible to formalize this set of verification in a checklist.

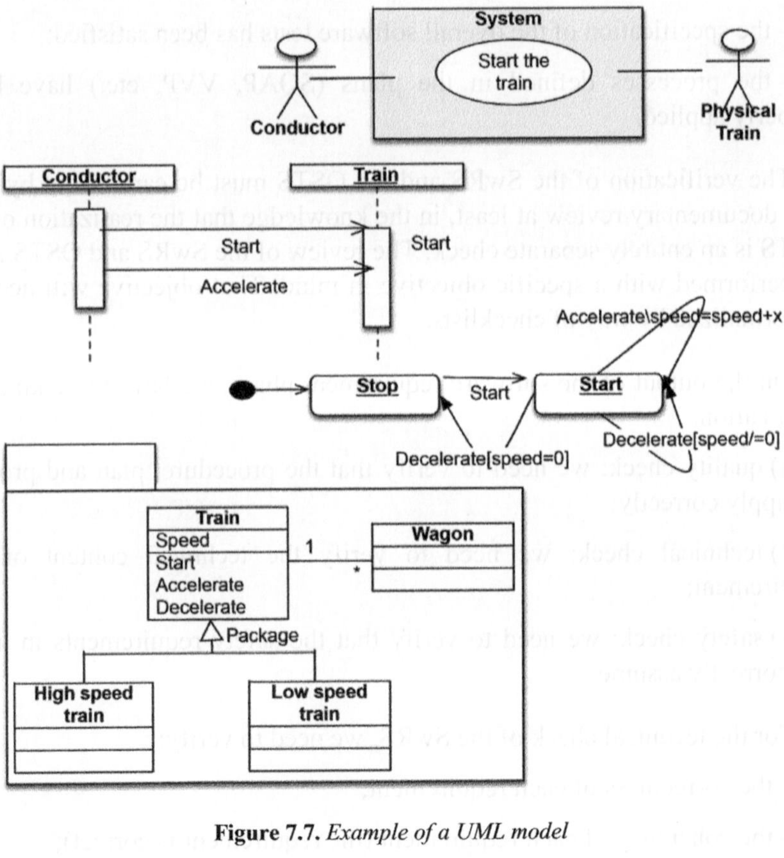

Figure 7.7. *Example of a UML model*

Checklists are used to guide the verifications on the industries (quality, V&V, safety, technical aspects) and on the hard points. Indeed, the checklists must take account of the company's accumulated experience. For example, if we use a UML model (see Figure 7.7, for example) as a support in the specification, we can identify points such as "For the overall model, the actors must be defined uniquely."

The verification report on the specification phase (SwRSVR) can thus be formalized as the written version of the end-of-phase review report, and may be produced in the wake of a formal review, referencing all of the records of the verification (list of input documents, completed checklists, review report, specific analyses, anomaly sheets list, etc.). The formal review has the aim of authorizing the beginning of the next phase (with or without reservations).

7.3.2. *Architecture and component design phase*

7.3.2.1. *Presentation of the phase*

The second phase of the V lifecycle for software realization is called the "architecture and design" phase. In actual fact, in involves two steps (see Figure 7.8): the definition of the architecture for the software and the decomposition of that architecture into its components. In addition, this phase of the lifecycle must enable us to identify the integration tests. The integration tests must cover two subjects: the software/software integration tests (which are used to verify the software interfaces) and the hardware/software integration tests (which focus on the verification of the hardware interfaces).

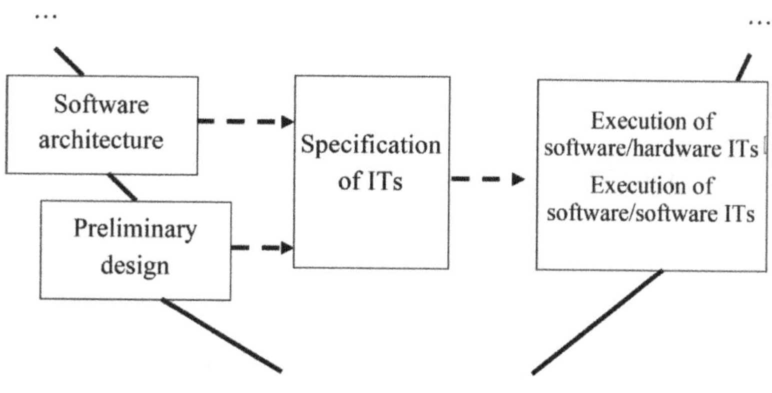

Figure 7.8. *Architecture and design phase*

As is shown by Figure 7.9, the input to the architecture and design phase includes the planning documents, the software requirements specification and methodological guides. On this basis, the first activity consists of describing the interfaces of the software being created; the second of identifying the software architecture; the third of performing the design; and to finish, we need to select the integration tests.

The phase concludes with the verification of all the productions (dotted arrows).

7.3.2.2. *Interfaces*

Figure 7.9 shows the first activity, which is the formalization of the interfaces. The new version of CENELEC 50128 includes the need to formalize the interfaces and produce a specific document, called the software interfaces specification.

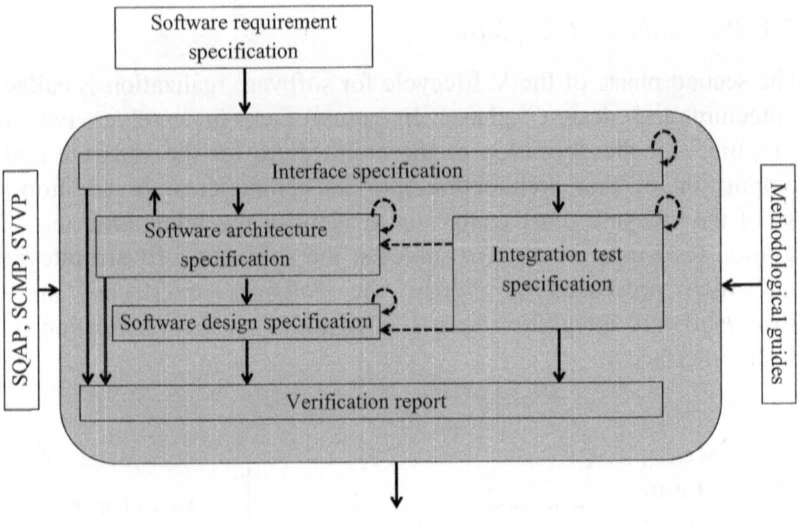

Figure 7.9. *Activity making up the architecture and design phase*

The identification of the interfaces was begun in the software requirements specification (see Figure 7.3). The document "software interface specification" must cover the description of the interfaces with the environment and the interfaces between the different software components (which is why there are two kinds of arrows in Figure 7.9).

The aim of the interface description is to cover external and internal interfaces and cover a number of topics:

– name;

– description;

– type: hardware, software;

– format: unit, precision;

– pre- and post-conditions;

– definition of equivalence classes (if they impact on the software's behavior), unacceptable values, unused values, maximum and minimum values, software behavior at these limit values and beyond these limits, etc.;

– time constraints;

– any exceptions;

– memory management: management of buffers and/or stacks, and the possible memory allocations associated therewith;

– the mechanisms of synchronization (such as scheduling);

– etc.

7.3.2.3. *Components*

In the 2001 version of CENELEC 50128, the basic unit is the module. A module is identified as (see Table A.20 and Definition B.30):

– having an entry point and an exit point;

– having a limited number of parameters;

– having a fully defined interface;

– being non-complex.

This modular approach puts one in mind of a function, a procedure, or a file containing a set of functions, and in a more evolved form, a class in an object-oriented language. It should be noted that in the 2001 version, there is a mixture of terminology, with the concepts of "subprogram" and "module" both being used. The concept of a module is not adequate to describe an architecture; indeed, we would like to be able to identify an element by a set perimeter in order to be able to replace it (in the same way we would a mechanical part) and re-use it.

If we seek to replace and re-use, we need to define the basic element of the architecture by means of a working perimeter and a set of interfaces. The working perimeter is characterized by a set of requirements and a set of functions. Additionally, in order for it to be reusable, the element's configuration needs to be carefully managed.

Techniques/measures	SSIL0	SSIL1; SSIL2	SSIL3; SSIL4
1. Information hiding	–	–	–
2. Information encapsulation	R	HR	HR
3. Parameter number limit	R	R	R
4. Fully defined interface	–	HR	M

Table 7.4. *Table A.20 from CENELEC 50128:2011*

The 2011 version of CENELEC 50128 introduces the concept of a software component, characterized by Definition 1.4. This definition conforms to Table A.20 of the standard (see Table 7.4).

Hereafter, therefore, the basic element is the component. Thus, we shall speak of component integration and component tests (rather than unit tests or module tests).

In the 2011 version of CENELEC 50128, Table A.20 (definition of a component) and Table A.3 (requirements for architecture) indicate that it is "highly recommended" (HR) to have encapsulation of the information, and that masking is not recommended (NR). Data masking poses a problem with regard to the observability of the software behavior, and therefore has an impact on testability and maintenance. Curative maintenance of a software application involves analysis of the behavior on the target machine in its real-world environment. Thus, it is necessary to have access to the software's internal statuses.

Encapsulation involves introducing data access services, which has a number of benefits:

– there is no longer direct access to the data, so those data are somewhat protected;

– it is no longer necessary to know the representation of the data in order to access it, which facilitates maintenance and reusability;

– etc.

Masking and encapsulation can be managed manually or be offered as a facility of the language (C++, ADA, etc.).

7.3.2.4. *Re-used components*

This section is based on section 7.3.4.7 of the new version of CENELEC 50128.

DEFINITION 7.1 (Re-used component).– *A re-used component is a component which has already been used. It is accompanied by documents demonstrating that a subset of its functions is guaranteed to conform to their requirements.*

7.3.2.4.1. Principles

There are two cases where components are re-used:

– the first case concerns the use of so-called COTS (commercial off-the-shelf) components, from outside the firm;

– the second pertains to the re-use of components previously developed and used. The new version of CENELEC 50128 makes use of the concept of pre-existing software; we shall in turn use the concept of pre-existing components.

In accordance with Definition 7.1, for a component to be able to be re-used, it must have:

– an identification and specification of the interfaces;

– an identification of the requirements which it respects;

– an identification of the hypotheses concerning the environment;

– an identification of the SSIL.

CENELEC 50128:2011 recommends that, for each re-used component, a fault study be performed with a view to analyzing the effects any such faults may have on the overall software.

7.3.2.4.2. Re-used components

For re-used components, it is possible to cater for the requirements set out in the previous section as regards the elements needing to be produced. This documentation (requirements, interfaces and hypotheses about the environment) can be compiled *a posteriori*.

For re-used components, it is possible to obtain detailed feedback on their previous use, which takes account of real-world use of one or more systems, and tells us about the duration of operation and number of hours of use. In accordance with Table E.9 – row 11 in CENELEC 50129:2003 (see Table 7.5), it is possible to construct a high level of confidence on the basis of the operational data.

Techniques/measures	SIL2 – SIL2	SIL3 – SIL4
High level of confidence demonstrated through use Optional when prior proof is not available	R: 10,000 hours of operating time and at least a year of experience in operation with the equipment in place	R: 1 million hours of operating time, at least two years of experience in operation with the different equipment in place, including safety analyses, with detailed documentation of the minor changes made during the time in operation

Table 7.5. *Table E.9 from CENELEC EN 50129:2003 [CEN 03]*

Re-used components must be identified (by name, reference and version), and we must demonstrate that the component is being used in a similar context to its previous uses (i.e. that the use context is appropriate for that component). The demonstration that the use context is similar must be based on an identification of the functions which are or are not used, on the use constraints and on the impact of any residual anomalies.

7.3.2.4.3. COTS

For COTS components (see Definition 1.5), it is rather difficult to obtain information about their specifications, their interfaces and the hypothesis about their environment. Therefore, the past experience and/or specific validation activities must be formalized in a COTS qualification report.

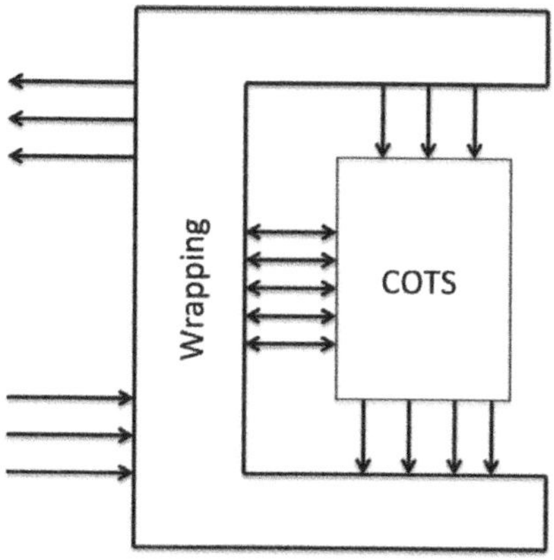

Figure 7.10. *Wrapping of a COTS component*

For each COTS component, we shall have a qualification report which indicates the usable functions, the associable safety integrity levels, the environmental hypothesis, the exported constraints and any residual anomalies. Thus, we must be able to show that the use of the COTS component in the project adheres to the constraints set in the qualification report.

The use of a COTS element must be carefully monitored in order to avoid the propagation of any faults to the components and devices which use that COTS element. In order to do so, it is possible to implement a strategy known as wrapping (see Figure 7.10).

7.3.2.4.4. Verification and validation

The re-used components must be integrated into the overall software validation, and we need to verify that the mechanisms to defend against any faults in the re-used components are effective and efficient.

This need must be taken into account when defining a process of verification and validation of the software under construction.

7.3.2.5. *Architecture*

Once the specification of the software application has been performed, it is then possible to put in place a defined architecture (see Figure 7.11). With this architecture, the software application is divided into components.

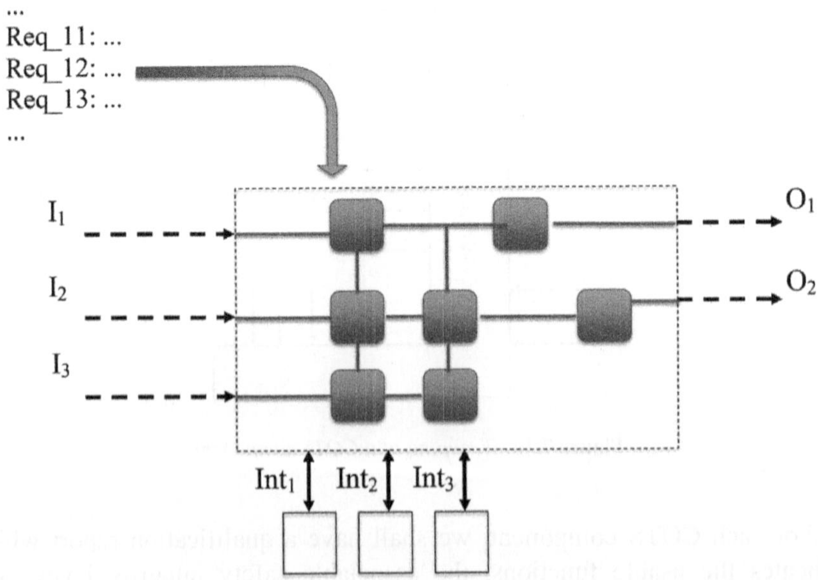

Figure 7.11. *From the requirements to a defined architecture*

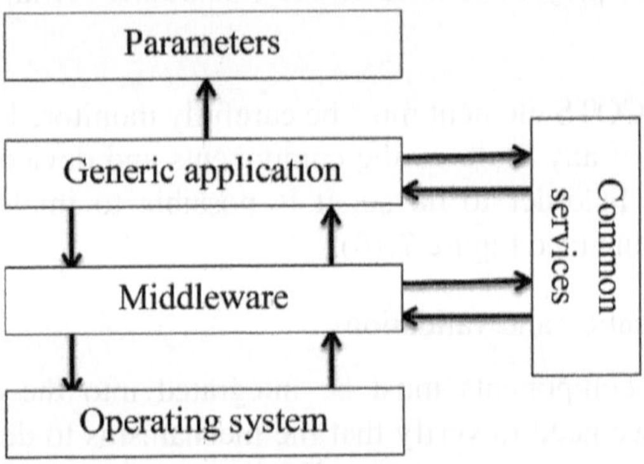

Figure 7.12. *Typical architecture*

Figure 7.12 shows a software architecture typical of the railway domain. We have an initial layer which is in charge of managing the hardware aspects and the execution (this could be an operating system, an execution loop, a sequencer, etc.). The second layer, known as *middleware*, handles communications. Thus, we have a generic application and its parameter set (see Chapter 6).

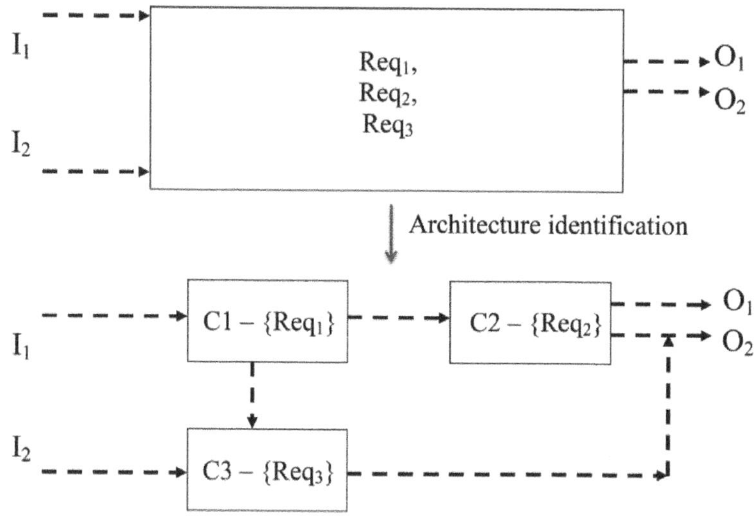

Figure 7.13. *Example of the architecture of a software application*

In Figure 7.13, on the basis of the requirements and the identification of the interfaces (see Figure 7.3), the software application has been divided into three components (C1, C2, C3). On the basis of the interfaces with the environment (I_1, I_2, O_1, O_2), the architecture identifies the components and the interfaces between them, and for each component the list of requirements to be taken into account is given.

The realization of the architecture must take account of criteria such as maintainability (ability to have a component replaced), re-use (capacity to re-use a component and definition of the criteria of re-use) and testability (we need to be capable of testing and running diagnostics).

Testability means that the behaviors must be observable. It is only possible to observe a behavior if it is not masked by another behavior. If the behaviors of the functions F and G are to alter the variable X, then the execution of the command "F; G" is not observable; indeed, the effect of the function F on the variable X is invisible, because the value of X is destroyed by the execution of G.

As regards the architecture rules, it is necessary to cover the following aspects:

– the complexity must be distributed throughout all the modules. If one or more of the modules are very complex (complexity is a concept which needs to be precisely defined: a high number of lines of code, many functions, a high cyclomatic number, etc.), we need to justify its inclusion as truly necessary;

– the number of components must be controlled;

– we need to avoid strong coupling between the components. When the components are strongly linked, the “plate of spaghetti” effect occurs. The alteration of a component impacts on all of the components;

– for each component, we need to identify whether it is new or re-used;

– we must carefully monitor the software safety integration levels between the components. Either we choose to realize the application at the highest SSIL identified, or we must implement architecture rules enabling us to partition the architecture and thus respecter the SSIL objectives of each component. The implementation of a partition or another approach requires us to demonstrate its efficacy and manage the problem of the exchanges between components with different SSILs;

– etc.

Technique/Measure	SSIL0	SSIL1; SSIL2	SSIL3; SSIL4
1. Defensive programming	–	HR	HR
2, 3, 4, 5, 6, 7, 8, 9, 10, 11, 12, 13, 14, 16			
15. Software Error Effects Analysis (SEEA)	–	R	HR
17. Information Hiding	–	–	–
18. Information encapsulation	R	HR	HR
19. Fully defined interface	HR	HR	M
20. Formal methods	R	R	HR
21. Modeling	R	R	HR
22. Structured methodology	R	HR	HR
23. Modeling supported by computer-aided design and specification tools	–	R	HR

Table 7.6. *Table A.3 from CENELEC 50128:2011*

CENELEC 50128:2011 identifies the resources needing to be implemented at the architectural level in its Table A.3 (see Table 7.6 here). This table can be broken down into five needs:

– rows 17, 18 and 19 relate to the definition of the components, and are actually a recap of Table A.20 (see Table 7.4);

– rows 20, 21 and 22 show the need to have at least a structured model. It should be noted that the need is more stringent than for the specification (see Table 7.3). Indeed, as can be seen from this section, an architecture requires a structured model (e.g. with boxes, interfaces or connectors). The aim of Row 23 is to encourage the use of tooled methods, with the objective being to improve reproducibility and verifications;

– rows 2, 3, 4, 5, 6, 7, 8, 9, 10, 11, 12, 13, 14 and 16 are safety implementation techniques applicable at device level. Indeed, it is no longer acceptable to introduce self-tests, diagnostics and/or reconfiguration into the software itself. The software is an object which is too complicated already to increase its complexity and thus damage its testability;

– row 1 relates to defensive programming. This is the most commonly mentioned technique to implement security in software, but it brings with it a number of disadvantages:

- it renders the code more complex – if we have a code whose complexity is n, the addition of a check by defensive programming at the start of the program will increase the complexity to $2n$,

- it increases the number of test cases needing to be performed,

- it introduces untestable execution paths into the software;

– row 15 pertains to software error effects analysis (SEEA). Note that this is the only analysis included in Table A.3, and that this analysis is designed to examine the impact of faults in the input data on a part of the software. In general, this analysis, which is employed here as a means of verification, is carried out by the team in charge of safety. At architecture level, SEEA can be used to identify the weak points (the need to add defensive programming), any conflicts relating to the SSIL (e.g. consumption of data with a different SSIL), etc.

In conclusion, the description of the software architecture comprises:

– a description of the boundary of the system (hardware interface, software interface);

– a model of the architecture;

– a list of components, with the list of requirements and the list of interfaces for each component;

– one or more traceability matrices.

7.3.2.6. *Software design*

7.3.2.6.1. Software design methods

The design method to be used to create the general design and to design the components must be selected on the basis of the SSIL needing to be attained.

Table A.4 (see Table 7.7 here) from CENELEC 50128:2011 can be broken down into four needs:

– rows 1, 2 and 3 pertain to the need to have at least a structured model (SSIL2) or a simulation (SSIL3-SSIL4). Note that the need is more stringent than it is for the architecture (see Table 7.6). Indeed, as can be seen from this section, an architecture requires a structured model (e.g. with boxes, interfaces or connectors), but we can go further with a simulation because we have data structures, sequencing information, etc.;

– rows 4, 5, 10, 12 and 13 are related to the programming paradigm put in place. CENELEC 50128:2011 recommends the putting in place of a component-based approach, but it must go hand in hand with a coding approach which respects the modularity, is based on a programming language, and may be procedural and/or object-oriented;

– rows 7 and 8 recap the fact that we need to choose a design methodology that enables us to detect anomalies as soon as possible (by typing) and analysis. It is possible to strengthen these properties by defining a subset (row 11) and/or design rules and coding rules;

– row 14 relates to meta-programming – it is not usable for the development of a railway application, but may be usable for a code generator or for data production.

The methodology of software design must therefore be described (modeling guide, design guide and coding rules) and it must be justified. The justification must be included in the SQAP or in a specific document.

Measure	SSIL0	SSIL1; SSIL2	SSIL3; SSIL4
1. Formal methods	R	R	HR
2. Modeling	R	HR	HR
3. Structured methodology	R	HR	HR
4. Modular approach	HR	M	M
5. Components	HR	HR	HR
6. Design and coding standards	HR	HR	M
7. Analyzable programs	HR	HR	HR
8. Strongly-typed programming language	R	HR	HR
9. Structured programming	R	HR	HR
10. Programming language	R	HR	HR
11. Language subset	–	–	HR
12. Object-oriented programming	R	R	R
13. Procedural programming	R	HR	HR
14. Metaprogramming	R	R	R

Table 7.7. *Table A.4 from CENELEC 50128:2011*

During the description of the methodology, we must demonstrate that the objectives from the standard are taken into account – particularly those in the tables in Appendix A. The HR techniques/measures that are not taken into account must be justified, but on the other hand if a technique/measure is not in the tables, it is not necessarily precluded, and a justification of its efficacy and relevance in relation to the body of the standard and to the need must be produced.

7.3.2.6.2. Design specification

For input to the component design (Figure 7.14), we must therefore have a document describing the architecture (Software Architecture Specification – SAS). However, monitoring the design of the components involves monitoring the quality of the design. Therefore, it will be necessary to have a guide presenting a set of safety principles and a programming guide.

In the context of certifiable software applications, we need to bear in mind that there are various objectives which must be served, such as: testability, maintainability and the ability to perform analyses to demonstrate the safety of the application.

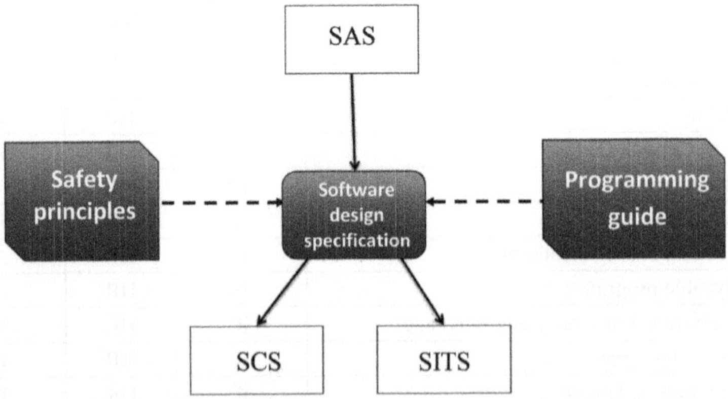

Figure 7.14. *Design process*

The ability to perform analyses on the design of the components means that this design must be clear and accurate. Everything must be documented in the SDS (software design specification).

The component design (Figure 7.15) is a step in the realization which involves finalizing the subdivision of the basic components. The basic components will depend on the design methodology used. They may be modules (in the sense of independent files), paquetage (in the sense of ADA [ANS 83, ISO 95, BAR 14] – they then consist of a specification part and a design part), classes and/or components (in the sense of the literature on components, which describes the component as being an independent autonomous entity).

Each basic unit (C_{ij}) is associated with a set of requirements written as $\{Req_{ij}\}$ and a set of functions/services/methods written as $\{F_{ij}\}$.

CENELEC 50128:2011 identifies the resources needing to be implemented at the level of the component design, in the form of Table A.4 (see Table 7.7 and the previous section).

The software component specification (SCS) must therefore identify the basic units C_{ij} and, for each C_{ij} describe the services and the requirements needing to be served. The SCS must show that all the components identified in the software architecture specification (SAS) have been taken into account, and that there is traceability between the requirements set out in the

SAS and those set out in the SCS. The component design file must identify all of the data structures, the sequencing and the error management mechanisms.

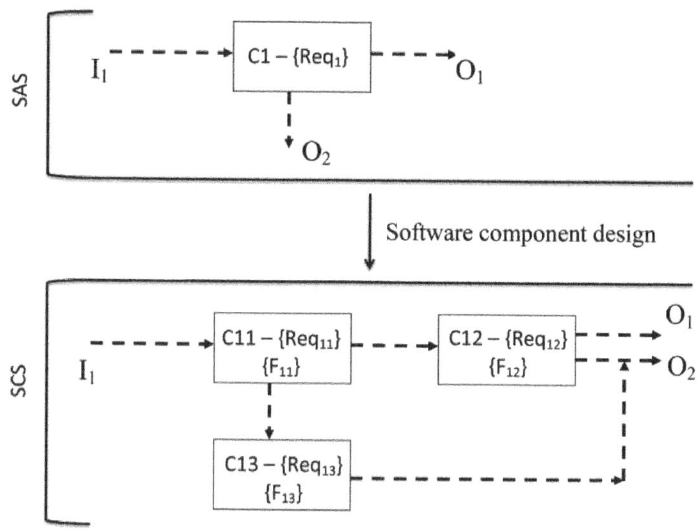

Figure 7.15. *Example of design*

To conclude, the description of the design of the components is made up of:

– a model of the architecture of each component;

– a list of components which the list of requirements and list of interfaces for each component;

– one or more traceability matrices.

7.3.2.6.3. Choice of development language

The software design phase is the time when we can choose the method used for the design of the components, and the programming language. Table A.12 from CENELEC 50128:2011 (reproduced below as Table 7.8) identifies metrics which the methodology must respect, but also represents a set of programming rules to be implemented. The programming guide must identify the programming rules, and these rules must be explained – it is important to understand the purpose of each of the rules, and the impact that non-respect of each rule would have on the product.

Techniques/Measures	SSIL0	SSIL1; SSIL2	SSIL3; SSIL4
1. Coding standard	HR	HR	M
2. Coding style guide	HR	HR	HR
3. No dynamic objects	–	R	HR
4. No dynamic variables	–	R	HR
5. Limited use of pointers	–	R	R
6. Limited use of recursion	–	R	HR
7. No unconditional branches	–	HR	HR
8. Limited size and complexity of functions, subroutines and methods	HR	HR	HR
9. Entry/exit points strategy for functions, subroutines and methods	R	HR	HR
10. Limited number of subroutine parameters	R	R	R
11. Limited use of global variables	HR	HR	M

Table 7.8. *Table A.12 from CENELEC 50128:2011*

Row 12 of Table A.4 indicates that we can use object-oriented programming; however, Table A.12 (rows 3, 4 and 5) introduces constraints pertaining to memory management which may not be compatible with object-oriented programming. The implementation of an object-oriented approach must entail the definition of a methodology and the demonstration that this methodology is acceptable in terms of the objective of the SSIL. The CENELEC 50128:2011 standard gives two new tables (A.22 and A.23) which introduce a number of measures for the realization of an object-oriented architecture and design. Tables A.22 and A.23 are, at present, insufficient to define a methodology.

Table A.12 in CENELEC 50128:2011 has been improved in relation to a number of different points:

– row 8 speaks of the need to keep the complexity of the code under control;

– row 9 refers to the fact that we must keep track of the exit points from a unit of code: the language C allows the execution of the command *return* more or less anywhere, which introduces an exit point and increases the cyclomatic complexity (Vg) by 1 for each *return* command. The addition of an exit point tends to introduce non-testable paths;

– row 10 is redundant with row 3 of Table A.20, which relates to the components;

– row 11 is important, and aims to prevent the program resembling a plate of spaghetti (the code cannot be separated into parts). The connection must be formed between this rule and the need for encapsulation cited in Table A.20.

The conventional process of development of a software application is based on the use of a programming language such as Ada, C and/or C++, for example. Although these languages have a certain level of abstraction in relation to the code executed on the final computer, they require the manual writing of lines of code, which contributes to the introduction of mistakes into that code (by human error).

Table A.15 from CENELEC 50128:2001 took account of the experience gained in this field in the early 2000s, so C and C++ were unacceptable without restriction. We can also see that for SSIL3-4, the use of Ada is merely "R", but becomes "HR" with the definition of a subset.

Measure	SSIL0	SSIL1; SSIL2	SSIL3; SSIL4
Ada	R	HR	R
MODULA-2	R	HR	R
PASCAL	R	HR	R
C or C++ with no restriction	R	–	NR
Subset of C or C++	R	R	R

Table 7.9. *Table A.15 from CENELEC EN 50128:2001*

Figure 7.16 shows a fragment of code in C which can generate two different codes, depending on the anomaly (a) or (b) which is implemented. This example highlights the weaknesses inherent in C; small programming mistakes are not detected at compile time. It should be noted that this type of error is detected if the programming language used is Ada [ANS 83, ISO 95, BAR 14].

It is possible to work around some of the shortcomings of C by putting programming rules in place. For instance, in order to prevent an anomaly such as that shown in Figure 7.16(b), we can implement a rule in the following form: "When comparing with a variable, that variable must be on the left-hand side of the expression".

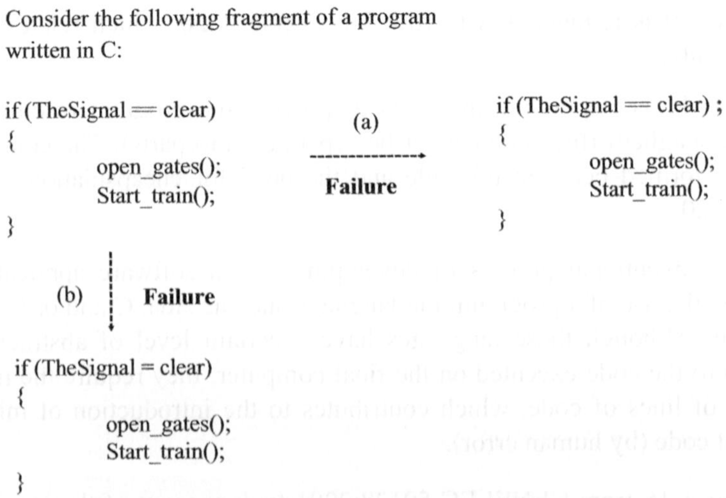

Figure 7.16. *Example of a fault in a program written in C*

Hence, the languages ADA and C are only used for a subset, so the new version of Table A.15 was produced. For C, the standard MISRA-C:2004 [MIS 04] is considered to be the best subset in the various domains. MISRA-C was first published in 1998 [MIS 98] and recently updated, in 2012 [MIS 12], which demonstrates a certain degree of maturity. With regard to C++, the situation is somewhat more delicate, because the earliest version dates from 2008 [MIS 08]. For this reason, we need to take account of a different standard, known as JSF++ [LM 05].

Measure	SSIL0	SSIL1; SSIL2	SSIL3; SSIL4
ADA	R	HR	HR
MODULA-2	R	HR	HR
PASCAL	R	HR	HR
C or C++ with no restriction	R	R	R
C#, JAVA	R	R	R

Table 7.10. *Table A.15 from CENELEC EN 50128:2011*

Table A.15 in CENELEC 50128:2011 includes new languages, such as C# and JAVA. These languages are merely "recommended" (but not "highly") and the difficulty lies in managing the execution (virtual machine), ensuring the quality of the code (with coding rules, limited complexity, etc.), the testing process and the capacity to measure the coverage of the tests.

It is, of course, possible to choose a language which is not included in Table A.15, but in that case, we need to demonstrate that the chosen language conforms to the requirements of the standard, that it is appropriate for the application being created, and it is necessary to analyze the language in order to identify any dangerous constructs which may require particular attention.

The choice of a language must be justified in the SQAP, and must be accompanied by the implementation of a coding guide and a set of programming rules.

7.3.2.6.4. Programming rules

The various standards recommend that programming and coding rules be formalized in a guide. Table 7.11 offers an example of a table that can be used to describe the programming and coding rules.

Name of rule	A unique identifier for each rule
SIL value	Each rule can be associated with one or more software security integrity levels: SSIL0/SSIL2/SSIL4
Description	The description of the rule may be the text used in the standard
Explanation	The description may be very brief, so an explanation may be needed to facilitate understanding
Examples	For each rule, it is necessary to introduce at least one example of correct use, and a counter-example to demonstrate incorrect use
Type	The aim of this attribute is to indicate whether the rules is "*Required*" or "*Advisory*" It is possible to define additional values to identify rules that have an impact on the form or on other aspects
Verification	Type of verification to be put in place: walkthrough, tools, etc.
Traceability	In this section, we refer to the various standards: - MISRA-C, - MISRA-C++, - CENELEC 50128, - JSF++, - etc.
Impact	This section contains an indication of the impact which could result from non-respect of the rule: e.g. impact on safety, impact on testability, impact on verification, impact on maintainability, etc.

Table 7.11. *Example of description of rules*

As introduced at various points in this chapter, the programming and coding rules must cover a variety of areas:

– formatting rules;

– good practice definition rules;

– language subset definition rules;

– etc.

7.3.2.6.5. Verification of programming rules

The definition of a subset of the programming language (section 7.3.2.6.3) and the definition of a set of coding rules (section 7.3.2.6.4) represent an initial step which must be accompanied by verification that the code actually respects the aforementioned rules.

A verification that the code respects the rules which have been defined can be performed manually if the number of rules is limited and/or the size of the code is small. It should not be forgotten that this verification must be auditable, which means that there is a formal record of the verification. In order to ensure the maintenance of the software application, this verification must be reproducible.

Therefore, it is preferable for this type of verification to be tooled. Therefore, the tool must be capable of analyzing the code for the selected language and it must be possible to define the rules to be verified. As we saw in the previous sections, there are recognized sets of rules such as MISRA-C [MIS 98, MIS 04, MIS 12], but we must also be capable of defining our own rules.

For example, Figure 7.17 is a screenshot taken during the execution of the QAC application on a program P written in C.

7.3.2.7. *Specification of the integration tests*

Based on the architecture, the component design and interface specification, it is possible to prepare the files for the integration tests. As the only input elements, at this point of the project, are descriptive documents (SAS, SDS, interface specification), we are specifying black-box tests (without knowledge of the realization).

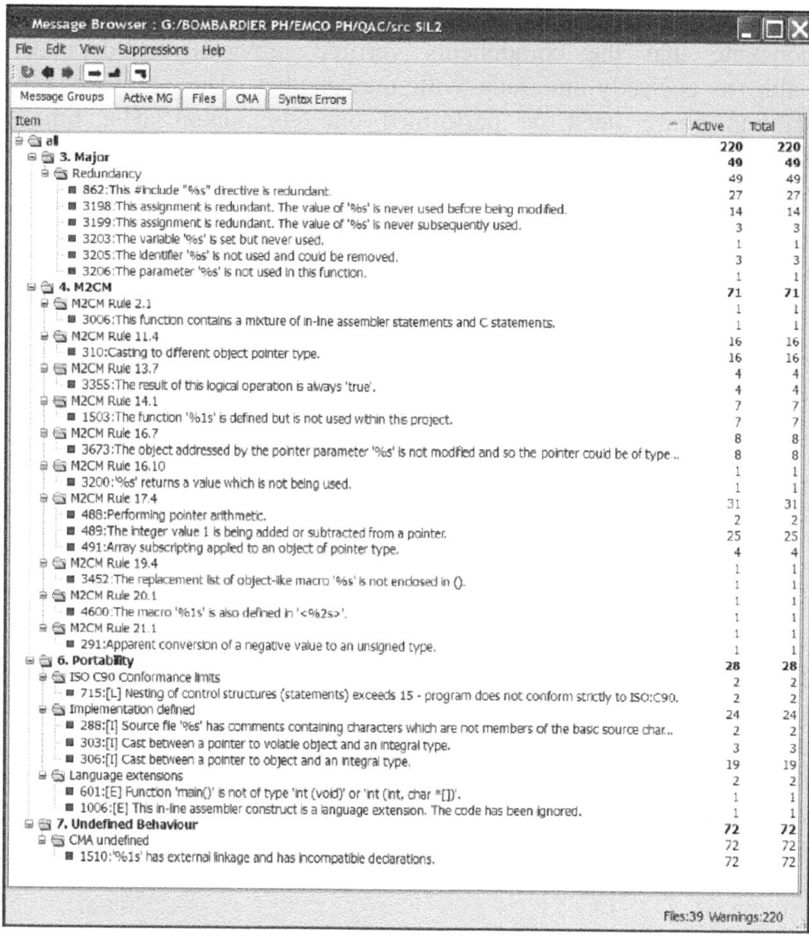

Figure 7.17. *Screenshot of the QAC summary screen*

As regards the integration tests, there are two groups of tests that need to be performed:

– software/hardware integration tests: this group of tests is designed to demonstrate that the software is capable of correctly handling the hardware interfaces. These tests may be performed on the basis of the components in the SDS or the SAS. The category may include tests pertaining to the satisfaction of the performance requirements (memory consumption, processor load, network load, etc.);

– software/software integration tests: this group of tests is designed to demonstrate that the components verified previously (independently) are able to correctly interface with one another. The software/software integration phase must be performed one component at a time (see Figure 7.18), until the overall software suite is obtained.

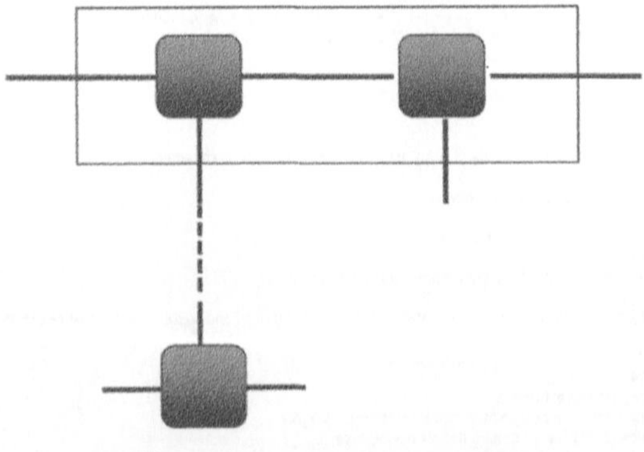

Figure 7.18. *Example of integration*

It is possible to draw up two different documents (SITS, HSITS) or only one (SITS) – in the latter case we speak of the SITS (Software Integration Test Specification). The purpose of the SITS is to demonstrate that the software/software and hardware/software interfaces are correct. The strategy used for the integration tests must be based on the overall methodology. Figure 7.19 illustrates the link between the preparation of the SITFs and the realization of the ITs.

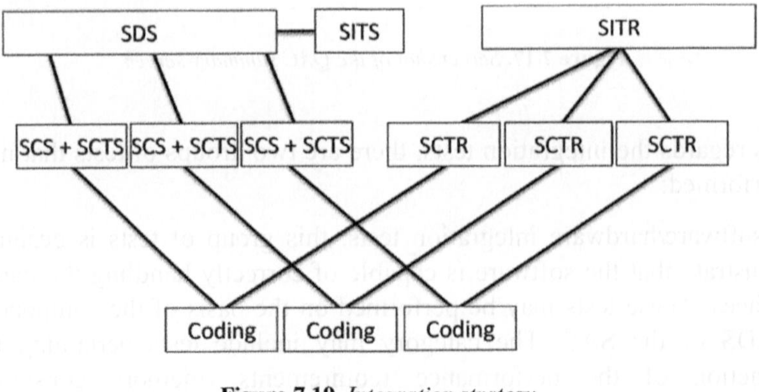

Figure 7.19. *Integration strategy*

The compilation of the SITS helps to demonstrate that the interfaces are testable – indeed, the person in charge of producing the SITS must (see section 5.7.2.4) carry out a fine-grained analysis of the interfaces (identification of min/max values for each input, identification of equivalence classes and identification of the observables).

On the basis of the interfaces, we identify test cases (TC_x). These test cases describe a particular situation to be attained, with this situation being linked to an equivalence class. Based on the test cases, it is possible to prepare the test scenarios; a test scenario describes a situation. Thus, a test case may play a part in multiple scenarios. The groundwork for the scenarios must be prepared in the SITF, and there is no need to actually have a computer file at this stage in the project.

The purpose of the SITS is to cover 100% of the interfaces. For each untestable interface, we must have a justification and an alternative verification activity. The SITS must contain a traceability matrix demonstrating that the interfaces have properly been tested and/or verified, but it is also necessary to demonstrate that each of the test cases is associated with an interface.

At the level of the SITS, it is possible to export tests to the overall software testing level. Indeed, installing the whole software package on its target platform may be seen as integration, but exported tests must be identified and justified. Similarly, it is possible to export integration tests to the component testing campaign – it may be the case that because of lack of observers, it is easier to carry out a component test, but this type of case causes problems in terms of testability and diagnostics for the final application.

Hence, the realization of the SITS is a verification of the description documents (SAS, SDS, interface specification).

7.3.2.8. *Verification report*

As indicated in section 3.6.1.3, CENELEC 50128:2011 identifies a verification report for the end of each phase. This verification report should demonstrate that:

– the software architecture specification has been satisfied;

– the software components specification has been satisfied;

– the software/software integration tests specification has been satisfied;

– the hardware/software integration tests specification has been satisfied;

– the processes defined in the plans (SQAP, VVP, etc.) have been correctly applied.

The SAS and SDS and of the SITS must be verified, at the very least, by a document review, given that the realization of the SITS is an entirely separate verification. The reviewing of the SAS, SDS and SITS must have a definite objective, which needs to be formalized in the form of checklists.

Hence, the software architecture and design phase verification report (SADVR) can be formalized as the result of the end-of-phase review, and be produced following a formal review referring to all the verification records (list of input documents, completed checklists, review report, specific analyses, list of anomaly sheets, etc.). The purpose of the formal review is to determine whether the project can move on to the next phase (with or without reservations).

7.3.3. *Component design phase*

As the input to the component design phase (see Figure 7.20), we have the software architecture and the software design, but also methodological documents such as a programming guide and/or a guide (or guides) describing the principles of safety implementation in a software application and the design method.

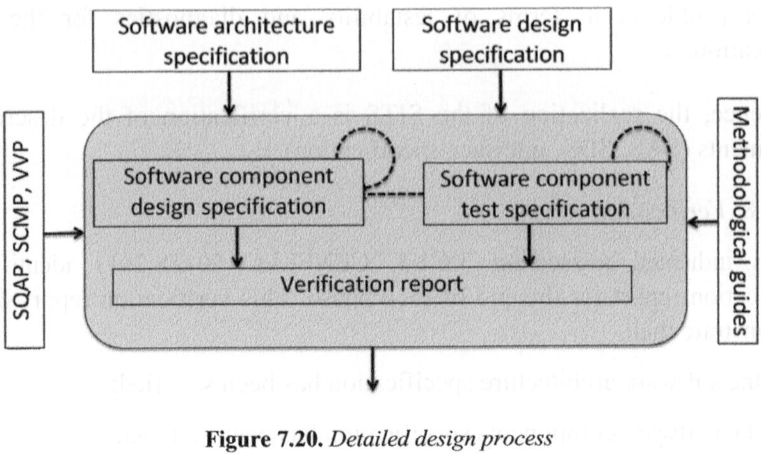

Figure 7.20. *Detailed design process*

7.3.3.1. *Realization of the design*

The objective of the component design is to describe the behavior of each base element (function or procedure). As is shown by Figure 7.21, the SCDS (Software Component Design Specification) contains, for each base unit (C_{ij}), a set of functions/services/methods represented as $\{F_{ij}\}$ which must be implemented.

The aim of the detailed design phase is to identify, for each service (F_{ij}), the algorithms and data structures. The algorithms must conform to the needs identified by the associated requirements (Req_{ij}).

The construction of these algorithms may necessity the implementation of additional services within that component.

In Figure 7.22, we can see how the function F13 in component C13 has been subdivided into three functions: F131, F132 and F133. The subfunction F131 is based on an internal service F1311, and all three – F131, F132 and F133 – rely on two internal functions: F1300 and F1301. We have also introduced the fact that all the services of component C13 can be based on one or more libraries. Figure 7.22 represents the call graph for function F131.

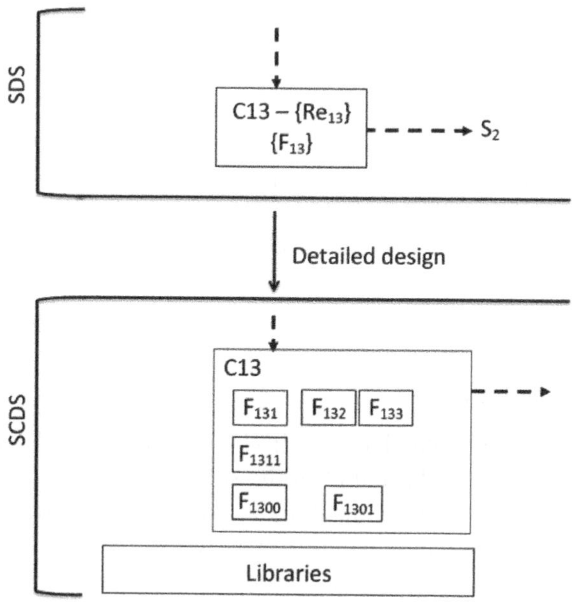

Figure 7.21. *Detailed design*

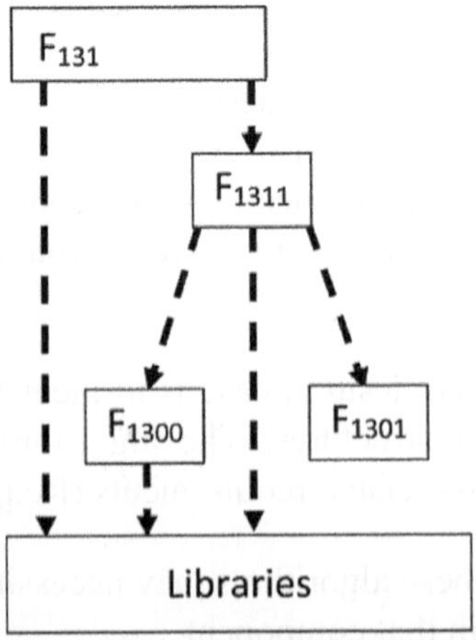

Figure 7.22. *Call graph for functions F131*

As regards the properties of the documents relating to the components, there is clarity (readability, comprehensibility, etc.) and particularly the capacity to be testable. Testability involves allowing for the observers and avoiding certain programming traits such as global variables, modifying input parameters, multiple exit points, etc., which is the reason for the methodological guide at the start of the phase.

7.3.3.2. *COTS and re-used components*

The preliminary and detailed design phases can be based on components which have already been used on other projects and/or COTS components.

We shall come back to the issue of re-used components and COTS in section 7.3.2.4.

7.3.3.3. *Component tests*

7.3.3.3.1. Methodology

It must be remembered that in CENELEC 50128:2001 (as with many equivalent standards), the base unit of the software was the module, with the concept of a module being linked to a fragment of code, a function, a procedure, a file, a class, etc., which is why we use the terms "unit test" and "modular test". The 2011 version introduces the concept of a component (see section 7.3.2.3) and therefore the concept of component tests.

The component test strategy must take account of the following objectives:

– each component must be tested;

– the tests used are black-box tests (with no knowledge of the code);

– the tests must enable us to show that the component does what is required of it by the component design;

– all parts of the component must be tested.

CENELEC 50128:2011 requires that the objectives given in Table A.5 (see the discussion in section 4.7.2.2) be respected.

The component test specification document must identify all component test cases. A test case comprises two parts:

– the input dataset, which is the set of values characterizing the initial situation (i.e. the inputs of the component/path in the algorithm);

– the expected outputs, which describe all of the output values in line with the input values (i.e. outputs from the component/calculation of the algorithm).

Figure 7.23 is an example of a test sheet. The part of the sheet in gray identifies the test cases, and the blank part of the sheet will need to be filled in when the tests are conducted.

The structure of the component tests specification document obeys the following plan:

– identification of the input elements;

– identification of the applicable plans (SQAP, SCTP[2], etc.);

– definition of the objectives (coverage, etc.);

– list of test cases;

– demonstration of coverage;

– balance of problems identified.

---

2 If needed, the component tests strategy can be described in the Software Component Test Plan.

Test name:	Name of the scenario		
Test Family: (if some family are define)	Family of the scenario.		
Upstream traceability: (What is tested)	List of requirement		
Description:	Description of the test with definition of the objective		
Pre-conditions:	Description of initial conditions.		Ok/ko
Required configuration:	Complete, partial, ...		Ref, version, …
ACTION:		Expected Result:	Result
Step number	Action description or description of a simulated specific state, generating outputs data to be analyzed	List of data to be checked at the current step, with the expected results.	Ok/ko (data, file, screen copy, photo, etc.)
…	…	…	
Tester name			xxx
Date			Xx/yy/aaaa
Global Result			OK KO with minor defect KO with major defect
List of CR			Xx, yy

Figure 7.23. *Example of test sheet*

If it is not possible to define a test case for certain requirements and/or portions of algorithm, it will be necessary to produce a justification, showing that this is an acceptable situation, and to put forward an alternative means of verification (see section 4.7.2.2 on verification).

7.3.3.3.2. Objective of coverage

In order to conform to the objectives identified earlier on, we need to define a strategy by which to select the test cases, which will depend on the methodology of specification of the components. This methodology must identify an objective of coverage of the tests in relation to the code. The 2001 version of CENELEC 50128 identified only the objective of coverage of the instructions, which is obviously insufficient, because there is no verification of the impact of the so-called implicit branches (e.g. an *else* branch which is not specified by an *if* structure).

The 2011 version of the standard introduces a new Table A.21 (see Table 7.12 here), which identifies different objectives in terms of coverage. The standard supplements this table by indicating that at component level, and for an SSIL3-4, we must go to the combinations 2+3, 2+4 or 5. The combinations 2+3 and 5 are similar, because in both cases, we test all the branches and all the conditions, but with the solution 2+3, less account is taken of the functional aspect. The combination 2+4 is more surprising, because we then take account of the usage of the data.

Measure	SSIL0	SSIL1; SSIL2	SSIL3; SSIL4
1. Statement	R	HR	HR
2. Branch	–	R	HR
3. Compound conditions	–	R	HR
4. Data flow	–	R	HR
5. Path	–	R	HR

Table 7.12. *Table A.21 from CENELEC EN 50128:2011*

CENELEC 50128:2011, for an SSIL2, only stipulates that the instructions must be met, which is insufficient – at the very least, it is crucial to verify the coverage of the branches.

For further information on this topic, readers can refer to section 4.7.2.4.2, which is devoted specifically to the coverage of the tests.

7.3.3.4. *Verification report*

As indicated in section 3.6.1.3, CENELEC 50128:2011 identifies a verification report for the end of each phase. This verification report must be able to demonstrate that:

– the software component design specification has been satisfied;

– the component test specification has been satisfied;

– the processes defined in the plans (SQAP, VVP, etc.) have been correctly applied.

The component design document and the component test specifications must be verified, at least, by a documentary review. The realization of the component test specification document is an entirely separate verification from the design document. The review must be performed with a definite objective, which needs to be formalized by way of checklists.

Hence, the software component design verification report (SCDVR) can be formalized as the result of the end-of-phase review, and be produced following a formal review referring to all the verification records (list of input documents, completed checklists, review report, specific analyses, list of anomaly sheets, etc.). The purpose of the formal review is to determine whether the project can move on to the next phase (with or without reservations).

7.3.4. *Coding phase*

The activity of coding consists of transforming a vision from components and algorithms into source code, with the constraint of having to respect various methodological guides, and the objective of doing only what is required. If there are difficulties during the coding which mean that something needs to be added to the design documents, the person in charge of the coding must report this, and any decision that is made needs to be formalized.

The activity of coding may be performed manually or automatically (see Figure 7.24). In the context of a manual coding activity, we must put in place a certain number of rules to guarantee that all of the needs outlined in the design are indeed taken into account (traceability), that the code will truly be testable and maintainable (readability, clarity, etc.). If an automatic code-generation process is used, we must demonstrate that the tools employed are consistent with the safety level to needing to be attained. It should be noted that very often, developers will use hybrid processes, whereby manual code is interfaced with automatically-generated code. In such cases, the interfacing must be completely and carefully handled.

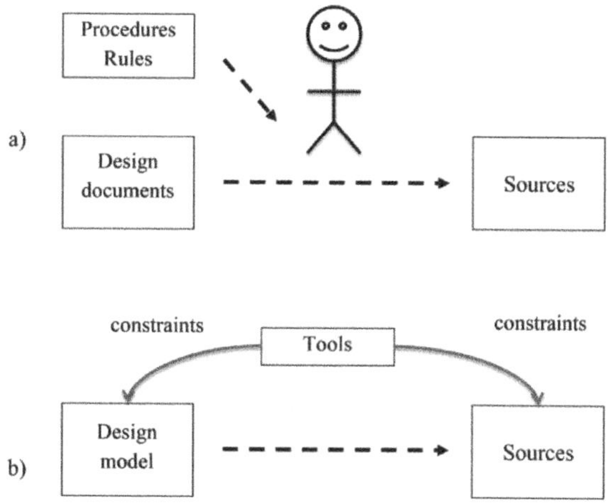

Figure 7.24. *Two examples of generation processes*

During the coding phase, the executable-generation sequence needs to be put in place. The executable-generation sequence may be a process of greater or lesser complexity. Figure 7.25 shows two examples of an executable-generation sequence. Version (a) is a linear sequence where a single executable file is produced from the sources, unlike version (b), which illustrates the production of two executables.

The source code must respect certain criteria. It must:

– be readable and comprehensible;

– be documented, and the comments must be useful (no description of the code itself, etc.);

– be managed in terms of the configuration;

– respect the methodological guides (naming rules, coding rules, constraints on the compilation sequence, etc.).

7.3.5. *Execution of component tests*

7.3.5.1. *Activity*

Figure 7.26 illustrates the activities connected to this phase. It is important to verify that the code conforms to its design and to the rules which have been formalized in the methodological guides.

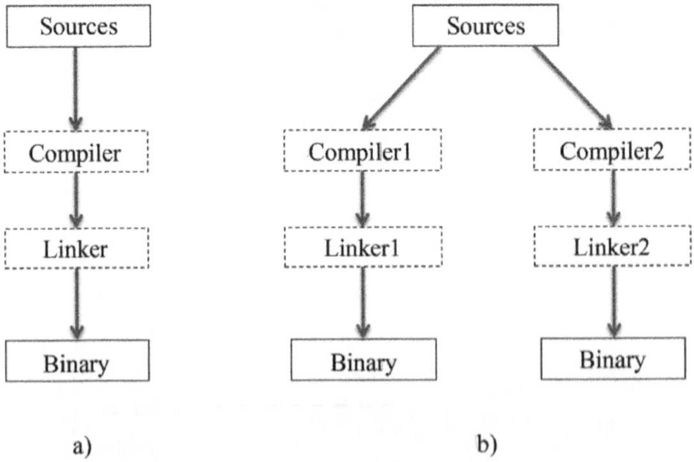

Figure 7.25. *Two examples of a compilation sequence*

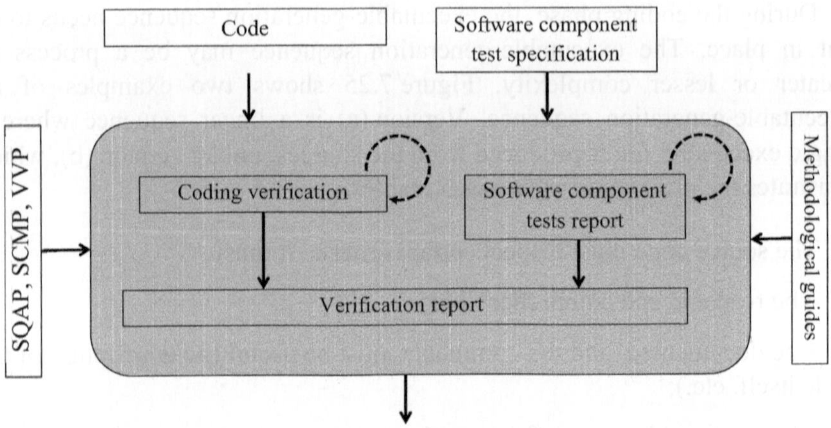

Figure 7.26. *Code verification and component test process*

The verification of the coding involves checking the quality of the code, staying on top of its complexity, respecting the methodological guides (naming rules, coding rules, compile-time sequence constraints, etc.).

The component tests (or unit tests or module tests) are carried out on a host machine (development machine) and generally do not require a test environment.

The carrying out of the component tests requires code fragments known as "stubs" to be put in place, in order to execute the piece of code we want to test. This "stub" is designed to be lost, because it is linked to a particular test and to a state of the code.

As is shown by Figure 7.27, the coding of pilot programs and fictitious modules (plugs) is part of the "stub". As a general rule, these elements are not delivered at the end of the process, and will not be configured. Therefore, it is impossible to conduct the component test in conditions mirroring those of the initial entry into service. In the case of a certifiable application, it is necessary to be able to repeat the component tests identically, and therefore we must closely document the testing process, manage the configuration of the input data and the results, and all the elements necessary for the execution of the unit tests (launcher, plugs, stub, etc.).

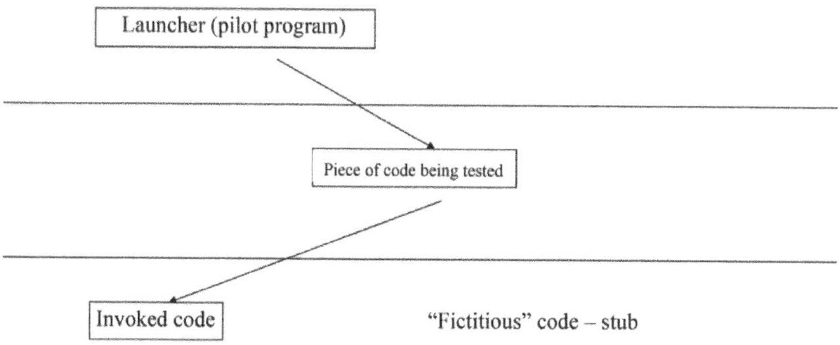

Figure 7.27. *"Scaffolding"*

Note that the component tests do not require us to have access to the target machine, and can be performed on the "host machine" (the machine used for development).

The software component test report needs to be to written at the end of the execution of those tests. It must enable us to identify what has been tested, by whom, when and the results obtained. This test report may be a specific document which repeats the test sheets or a supplemented copy of the software component test specification document. Figure 7.23 is an example of a sheet describing a test case.

The structure of the test report is as follows:

– identification of the input elements;

– identification of applicable plans (SQAP, etc.);

– resources used: people, tools, environment, etc.;

– version analyzed;

– list of test cases;

– demonstration that the objectives have been attained;

– result of measurement of the test coverage and justification;

– summary of problems identified;

– conclusion as to the correctness of the version.

The summary of the problems identified means that, for each detected fault, an anomaly sheet has to be created. This anomaly sheet must describe the identified problem and the activity and elements employed to remedy it.

7.3.5.2. *Verification report*

A verification must be performed on the generation sequence (see Figures 7.24 and 7.25) in order to verify that the generation of the executable works properly, and that it respects the safety objectives and the constraints exported by the various tools employed.

As with all the phases, a verification report linked to the source code needs to be produced. This report must summarize the various activities of verification (walkthrough, complexity measurement, checking of methodological guides, execution of test cases, etc.), list any problems identified (for each problem, an analysis/justification must be produced), and must draw a conclusion as to whether or not the code conforms to the design documents and the applicable guides.

7.3.6. *Software integration phase*

7.3.6.1. *Activities*

Figure 7.28 shows the three activities that need to be performed in this phase. On the basis of the code, we need to perform the integration of the components in order to obtain the final application. The software/software integration process is defined in the S/S integration tests specification – now we simply need to implement it. As indicated in section 4.7.2, the process is one of integration, one component at a time. It should be noted that for the S/S integration, we do not need to have access to the target machine, as the process can be performed on the machine used for the development (the host machine).

For H/S integration, we may need to use a specific strategy, because we need to have the target machine and certain devices in order to have access to data pertaining to the status of the hardware components.

7.3.6.2. *Verification report*

As with all phases, a verification report pertaining to the integration activities must be produced; this report must summarize and evaluate the various verification activities involved in the integration tests (generally a documentation walkthrough).

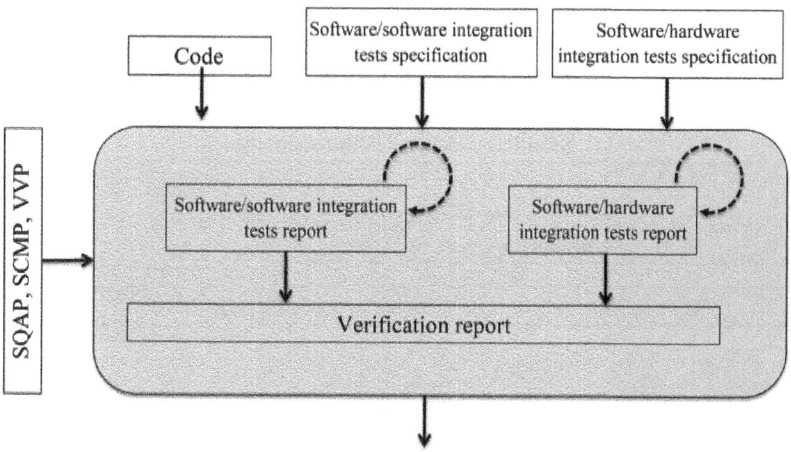

Figure 7.28. *Software integration process*

7.3.7. *Overall software testing phase*

Figure 7.29 shows the two activities involved in this phase – namely the performance of overall software tests on the target machine and the production of the software validation report.

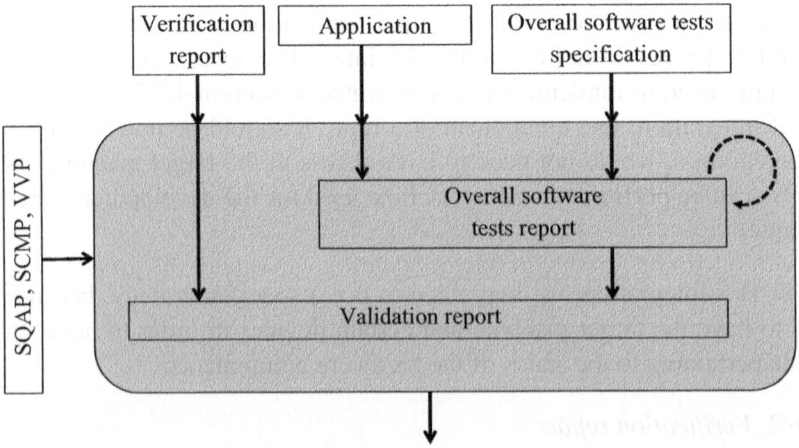

Figure 7.29. *Overall software testing process*

The overall software tests on the target machine enable us to demonstrate that the hardware/software ensemble does indeed conform to the specification of the software requirements on the final equipment. Figure 7.30 shows the overall software testing environment.

Figure 7.30. *Software testing on the target machine in a simulated environment*

As the overall test is directly linked to the program's execution, the validity of this activity is strictly dependent on the environmental conditions of the test. These conditions must, as faithfully as possible, reflect normal usage conditions – in other words, the operational environment.

The 2011 version of CENELEC EN 50128 alters the notion of "validator". This is a person who is responsible for making sure the overall tests identified from the requirements in the specification are carried out, but s/he may also request the addition of system-oriented tests and/or user-needs-oriented tests. The validation supervisor is responsible for the validation report.

The verification report for the previous phases is thus replaced by a software validation report (SVR), which takes account of all of the results obtained up until that point. Indeed, in order to compile the validation report, we take all of the verification reports that have been produced, we take the results of the overall tests (and the verification associated with those tests) and perform an analysis of the coverage of the objectives (relating to the requirements, the interfaces, the code, etc.). In addition, article 8.7.4.9 of the standard necessitates confirmation that Appendix A of CENELEC 50128:2011 has been properly applied (choice analysis and verification of justifications). The format of this verification report is described in article 8.7.4.10 of CENELEC 50128:2011.

The software validation report must make a judgment as to the adequacy of the software for its intended use. It may identify exported constraints. In addition, if the software is appropriate for the intended use, a delivery sheet must be drawn up, containing the usage constraints.

7.4. Some feedback on past experience

Chapter 7 of CENELEC 50128:2011 introduces a V-lifecycle for development (see Figure 7.1), where the tests (overall software tests, software integration tests and software component tests) take place during the descendant phase, which enables us to identify any *bugs* in the software as early as possible. By using this approach, we are able to launch into the ascendant phase with a minimal number of bugs. Indeed, what increases the cost and the time taken for the development is the process of bug management.

The cost of correction of a bug is directly linked to the phase during which that bug has been identified. If a bug is detected during the functional test phase, it costs between ten and a hundred times more (or even more than this, in some cases) to rectify than a bug identified during the component testing phase. This cost is linked to the resources which have been mobilized up until the point when the fault is detected, and to the difficult of performing functional tests (use of a target machine, the need to carry out the tests in real time, difficulty in observation, technical knowledge of the people involved, etc.).

The author's own experience (as an assessor/certifier of railway systems) has led him to conclude that the phases of component testing (be it module or unit testing) and integration testing are generally ineffective, given that industrial actors believe that:

– component tests (module/unit tests) are useless (as a general rule, the unit tests are defined on the basis of the code);

– software/software can be boiled down to a "big-bang integration" (integration of the whole code instead of integration one module at a time) – in the worst case scenario, the code is all compiled in one go, and the integration can be summarized as the verification of the interfaces by the compiler;

– the hardware/software integration is handled by the functional tests on the target machine. If the overall software runs correctly on the target machine, then the integration is correct.

As an assessor, the author is aware of the financial issue, but costs should not be reduced by cutting corners in terms of standard, but rather by strict use in combination with a body of experience.

7.5. Conclusion

In this chapter, we have presented the development of a generic application as defined in Chapter 7 of CENELEC 50128:2011. We have presented the V lifecycle and, for each step, the techniques, resources and elements needing to be produced and verified.

CENELEC 50128:2011 is based on a V lifecycle, but it brings back to center stage the concept of software assurance, which combines the necessary requirements: quality management, skill management, tool management, configuration management, V&V management and external evaluation. In order to do so, the proposed cycle contains the verifications, and they are performed as early on in the cycle as possible. This enables us to have what is known as an "agile" approach: we seek to discover any faults as early as possible, not by executing the software but instead by preparing tests.

This approach enables us to develop software applications at lesser costs and within a reasonable period of time if the verifications are efficient.

7.6. Appendix A: the programming language "Ada"

The earliest railway applications in France, in the mid-1980s, were programmed with the language Modula 2. However, afterwards, Ada 83 [ANS 83] became the reference language for the development of critical applications [RIC 94]. As is shown by Table 7.9, in applications with a high level of criticality (SSIL3/SSIL4), the language Ada itself is only "R" (Recommended). Thus, we need to put in place a subset of the language so that the use of Ada becomes HR.

The maturity of the language and knowledge of the subset mean that now, Ada is HR as a programming language in this context (see Table 7.10).

Ada was designed on the initiative of the DoD (the Department of Defense in the USA) to bring the 400+ languages or dialects used by that organization in the 1970s into one form.

Ada is very widely used in onboard software applications in avionics (the Airbus), the space domain (the Ariane rocket) and in the railway domain. The primary characteristic of these systems is that they require the execution to be absolutely flawless.

Ada 83 evolved into a second major standard: Ada 95 [ISO 95], which was the first standardized object-oriented language. It brings the possibility of constructing object-oriented models. The latest dated version is called Ada 2012 [BAR 14] and introduce new features such as properties description that offer an opportunity to introduce some formal verification.

```
with Ada.Text_IO;
procedure Hello is
begin Ada.Text_IO.Put_Line("Hello, world!");
end Hello;
```

Figure 7.31. *Example of code in Ada*

As regards the certification of Ada compilers, the existence of a standard and of fairly detailed set semantics for the language has enabled us to define a process for certifying a compiler. This process has been implemented on a variety of compilers. The process is based on a suite of tests called ACATS *Ada Conformity Assessment Test Suite* (ACATS) – see the standard [ISO 99a]. For further information, consult [ADA 01].

For the time being, these new versions of Ada have not been adopted in the field of onboard systems, because of the object-oriented aspect. However, in light of their effectiveness, Ada 95 compilers are used for compilation, in the knowledge that we are dealing with a subset of the language which does not use "object-oriented" traits. At present, ADA 2012 is not used for critical safety applications.

The "object-oriented" aspect is not taken into account by the standards applicable to critical applications – CENELEC EN 50128 [CEN 01, CEN 11], DO 178 [ARI 92, DO 12], IEC 61508 [IEC 08], ISO 26262 [ISO 11], and IEC 60880 [IEC 06].

In order to get around this pitfall, ISO 15942 [ISO 00] defines a restriction on the constructions of the Ada 95 language and usage rules (programming styles) which can be used to realize a so-called certifiable application.

The language SPARK Ada [BAR 03] is a programming language which is a subset of Ada. All the complex structures in Ada considered to be risky or not facilitating an easy demonstration of the safety are omitted from SPARK Ada. A mechanism that enables annotations to be added into the code has been put in place.

The SPARK Ada tools contain a compiler but also an annotating verifier. It should be noted that there is a free version of the SPARK Ada tools[3]. Chapter 3 of [BOU 12] presents the SPARK Ada tools and industrial examples of their implementation.

A new version of Spark Ada introduces the possibility to use the properties introduced in Ada 2012 to carry out some formal verification.

3 To learn more about AdaCore and open versions of tools such as GNAT and SPARK Ada, see the Website: www.libre.adacore.com.

7.7. Appendix B: the programming language "C"

7.7.1. *Introduction*

The language C[4] [KER 88] is one of the first languages to have been made available to developers to create complex applications. The main difficulty with C lies in the partial definition of the language, which means that different compilers generate an executable with different behaviors. It has since been subjected to a standardization process by the ANSI [ISO 99].

As regards the use of C [ISO 99], depending on the desired safety integrity level, CENELEC EN 50128 [CEN 01] recommends defining a subset of the language (see Table 7.9), whose execution can be controlled.

Table 7.10 shows that it was considered that sufficient experience had been gained in the languages Ada, C and C++, meaning that the concept of a subset of the language no longer needs to be explicitly mentioned, because it is considered as having been acquired.

7.7.2. *The difficulty with C*

Some of the weaknesses of C [KER 88, ISO 99] can be circumvented by putting programming rules in place. For example, in order to prevent an anomaly of the type *if* (a = cond) instead of *if* (a == cond), we can put in place a rule in the following form: "When comparing with a variable, that variable must be on the left-hand side of the expression". This gives us: *if* (cond == a).

From 1994, from the existing feedback concerning the use of C (for example, see [HAT 94]), it became clear that it was possible to define a subset of C that could be used to create software applications requiring a high level of safety (SSIL3-SSIL4).

In fact, for C, the norm MISRA-C [MIS 04] has become a *de facto* standard, which was developed by the *Motor Industry Software Reliability Association* (MISRA[5]).

---

4 Although Kerdigan and Ritchie [KER 88] do not describe the language C as understood by ANSI [ISO 99], theirs remains one of the most interesting books on the subject.
5 For further information, see: www.misra.org.uk/.

7.7.3. *MISRA-C*

MISRA-C [MIS 04] specifies programming rules (see examples given in Table 7.13) that help to avoid runtime errors caused by improperly-defined constructs, unexpected behavior (certain structures in C are not completely defined), misunderstandings between the people in charge of the realization (readable code, code with defaults, etc.). There are various tools which can be used to automatically check the MISRA-C rules.

ID	Status[6]	Description
Rule 1.1	Required	All of the code must conform to ISO 9899:1990 "Programming languages – C", amended and corrected by ISO/IEC9899/COR1:1995, ISO/IEC/9899/AMD1:1995 and ISO/IEC9899/COR2:1996
Rule 5.4	Required	Each tag must be a unique identifier
Rule 14.1	Required	There must be no dead code
Rule 14.4	Required	The must be no goto statements in the programs
Rule 14.7	Required	A function must have a single, unique exit point at the end of the function
Rule 17.1	Required	Pointer arithmetic may only be used for pointers that lead to a table or an element in a table.
Rule 17.5	Advisory	An object declaration should not contain more than two levels of indirection of the pointer

Table 7.13. *Some of the rules[7] set out in MISRA-C:2004 [MIS 04]*

The MISRA-C standard [MIS 04] repeats rules (e.g. see rules 14.4 and 14.7) which are explicitly given in a number of other standards:

– rule 14.4: in CENELEC 50128 – Table A.12, or IEC 61508 – Table B.1;

– rule 14.7: in CENELEC EN 50128 – Table A.20, or IEC 61508 – Table B.9;

– etc.

---

6 A MISRA rule may be "*required*" or "*advisory*". A "required" rule must be implemented by the developer, and an "advisory" rule must not be overlooked, even though it is not absolutely compulsory to implement it.

7 [MIS 04] introduces 122 "required" rules and 20 "advisory" rules.

MISRA-C [MIS 98] was developed in 1998, and updated in 2004 [MIS 04] and then later in 2012 [MIS 12], which demonstrates that there is a certain amount of experience that has been gained from feedback.

The main difficulty with C is choosing a compiler that has garnered enough experience for the chosen target and for the desired SSIL. In the absence of a precise and complete standard, there is no certification process for C compilers, even though initiatives such as [LER 09] are in place[8].

7.7.4. *Example of a rule*

Rule_x: all blocks must be marked with a beginning and an end.

The structure:

```
for (i=1; i<= length; i++)
        x = i + 3;
```

needs to be replaced by:

```
for (i=1; i<= length; i++)
    {x = i +3;}
```

7.8. Appendix C: introduction to object-oriented languages

As already stated earlier, the "object-oriented" aspect is not taken into account by the applicable standards – CENELEC EN 50128 [CEN 01], DO 178 [ARI 92, DO 12], IEC 61508 [IEC 08] and ISO 26262 [ISO 11] – for critical applications.

The object-oriented aspect is cited in CENELEC EN 50128 [CEN 01], but the constraints that apply to the languages mean that we cannot develop a

8 Note that [LER 09] shows the task of verification of a restricted C compiler, but that the term "certification" is introduced. This does not mean certification by an external organism. Certification does not relate to demonstrating that the compiler is right, but must also be applied to the process used, to the elements produced (documents, sources, etc.) and to the tools employed (Ocaml, Coq and other open tools must be demonstrated as being appropriate for use for creating safety-related software applications).

critical application (SSIL3 and SSIL4) with this type of language (see Table 7.14).

Technique/measure	SSIL0	SSIL1	SSIL2	SSIL3	SSIL4
No dynamic objects	–	R	R	HR	HR
No dynamic variables	–	R	R	HR	HR
Limited use of pointers	–	R	R	R	R

Table 7.14. *CENELEC EN 50128 [CEN 01] – Table A.12*

As is demonstrated by Tables 7.9 and 7.10, the language C++ [ISO 03, ISO 06] is cited as being applicable, but certain recommendations are not compatible with the use of an object-oriented language, as shown by Table 7.14.

C++ was developed in the 1980s as an improvement on C. C++ introduces the concept of class, of inheritance, of virtual functions and of overload. A standard was published for C++ by the ISO in 1998 and again in 2003 [ISO 03].

Since the beginning of the millennium, many projects have been undertaken in an attempt to define a framework to govern the use of C++ [ISO 03, ISO 06] for the development of applications with a high software safety integrity level (SSIL3-SSIL4).

For example, we can cite the work of:

– OOTiA[9] (*Object Oriented Technology in Aviation*), which has published numerous guides [OOT 04a, OOT 04b, OOT 04c, OOT 04d];

– JSF++ (*Joint Strike Fighter C++*), which published a guide [LM 05] regarding development in C++. This guide summarizes the pre-existing works, and particularly the MISRA-C standard [MIS 98];

– MISRA, which developed the MISRA-C++:2008 standard [MIS 08]. Table 8.12 presents examples of rules in MISRA-C++:2008.

9 For further information, see: www.faa.gov/aircraft/air_cert/design_approvals/air_software/ oot/.

Hence, C++ [ISO 03, ISO 06] is a fairly old language. Approaches identifying the weak points of C++ and suggesting rules began to appear quite early on [SCO 98, SUT 05], but the definition of a guiding framework for the use of C++ for high-SSIL applications is fairly new [LM 05, MIS 08, OOT 04], which explains why we find applications in C++ up to SSIL2.

As shown by Table 7.15, the MISRA-C++:2008 standard [MIS 08] introduces a certain number of rules inspired by the existing ones for C, but they are unable to take account of all the difficulties with C++.

Owing to a variety of pressures (such as the decrease in the number of Ada and C programmers, for instance), the updating of standards such as CENELEC EN 50128 or DO 178 has given rise to initiatives aimed at introducing object-oriented languages.

Id	Status[10]	Description
Rule 0-1-1	Required	A software application must not contain unreachable code
Rule 0-1-2	Required	A software application must not contain any non-executable paths
Rule 1-0-1	Required	All the code must conform to ISO/IEC 14882:2003 "The C++ standard incorporating Technical Corrigendum 1"
Rule 2-10-4	Required	The name of a "class", "union" or "enum" must be a unique identifier
Rule 5-2-3	Advisory	The operation "cast" from a basic class to a derived class must not be applied to polymorphic types
Rule 15-5-1	Required	A class destructor must not end with an exception
Rule 17-0-4	Document	All the libraries must conform to MISRA-C++
Rule 18-0-1	Required	The C library must not be used

Table 7.15. *A few rules[11] from MISRA-C++:2008 [MIS 08]*

10 A MISRA-C++ rule may be "*required*", "*advisory*" or "*document*". The implementation of a "required" rule by the developer is mandatory; an"advisory" rule cannot be ignored even though it is not obligatory to implement it; and a "document" rule is obligatory.

11 [MIS 08] introduces 198 rules.

Thus, the new version of CENELEC EN 50128 [CEN 11] extended the list of usable object-oriented languages to JAVA and C#, as shown by Table 7.10. However, this new version of the standard introduces a few restrictions (limitation on inheritance), shown by Table 7.16.

The version for C of the DO 178 standard has a specific appendix which defines the constraints of implementation of an object-oriented language for the realization of a critical application. This appendix is extremely restrictive, if not totally inapplicable.

As is the case with C, the difficult point with C++ remains the demonstration that the compiler, for the chosen target and the associated libraries, respects safety objectives which have been defined by the safety studies. As no certification is in place for C++ compilers, we need to put in place a justification based on gained experience and/or qualification.

	SSIL0	SSIL1	SSIL2	SSIL3	SSIL4
The classes should have only one objective	R	R	R	HR	HR
Inheritance used only if the derived class is a refinement of its basic class	R	HR	HR	HR	HR
Depth of inheritance limited by coding standards	R	R	R	HR	HR
Neutralization of operations (methods) under strict control	R	R	R	HR	HR
Multiple inheritance used solely for the interface classes	R	HR	HR	HR	HR
Inheritance from unknown classes	–	–	–	NR	NR

Table 7.16. *New CENELEC EN 50128 [CEN 11] – Table A.23*

7.9. Appendix D: documentation needing to be produced

Document title	Acronym
Specification	
Software Requirements Specification	SwRS
Overall Software Test Specification	OSTS
Software Requirements Verification Report	SwRVR

Document title	Acronym
Architecture and design	
Software Architecture Specification	SAS
Software Design Specification	SDS
Software Interface Specifications	SIS
Software Integration Test Specification	SITS
Hardware/Software Integration Test Specification	HSITS
Software Architecture and Design Verification Report	SADVR
Component design	
Software Component Specification	SCS
Software Component Test Specification	SCTS
Software Component Design Verification Report	SCDVR
Component implementation and testing	
Software Source Code and Supporting Documentation	–
Software Component Testing Report	SCTR
Software Source Code Verification Report	SSCVR
Integration	
Software Integration Test Report	SITR
Hardware/Software Integration Test Report	HSITR
Software Integration Verification Report	SIVR
Overall software tests	
Overall Software Test Report	OSTR
Software Validation Report	SVR
Tool Validation Report	TVR
Delivery Sheet (Software Version Sheet)	SwVS

8

Modeling and Formalization

8.1. Introduction

In this chapter, we shall present various concepts such as models, formal methods and formal techniques. We shall place emphasis on the advantages of these approaches and techniques to realize a functionally safe software application.

As we have demonstrated in the collection [BOU 11a, BOU 12a, BOU 12b, BOU 14c], by means of concrete and complex projects, the railway domain has experienced the use of formal methods (Z [SPI 89], B-method [ABR 96] and SCADE [DOR 08]) and of formal techniques (proof, model-checking and abstract interpretation of programs).

From these various experiments, it has emerged that modeling is a fundamental need for the different phases of realization of a software application. The main evolution in this field in recent years is the recognition of the necessity of creating models.

8.2. Modeling

8.2.1. *Objectives*

In order to work on the correctness and completeness of the requirements, it is helpful to create one or more models. A model is an abstract description of a system and/or a process. As it is a simplified representation, there is a reduction in complexity, the model can be easily manipulated and helps the thought process and/or the performance of a certain number of verifications. The more accurate the model, the closer the results obtained will be to those which would actually be observed on the finished system.

Therefore, the model can serve as a tool for communication between the various stakeholders.

A good model should focus on the problem, but it may go so far as to describe a solution, and it is then possible to manage all or part of the final code. Thus, a model may or may not be "executable".

More generally, it is possible to have models devoted to each step in the process of realization of the system; we then speak of *Model-Based Development* (MBD). In the author's native France, we speak of IdM – "*Ingénierie des Modèles*".

Modeling must be used in addition to the textual description of the requirements – we must be able to express our need in order to model it. To implement a model without a textual requirement is tantamount to exhibiting one's needs without having expressed them, and the question then arises of how to check that the model does actually correspond to the need.

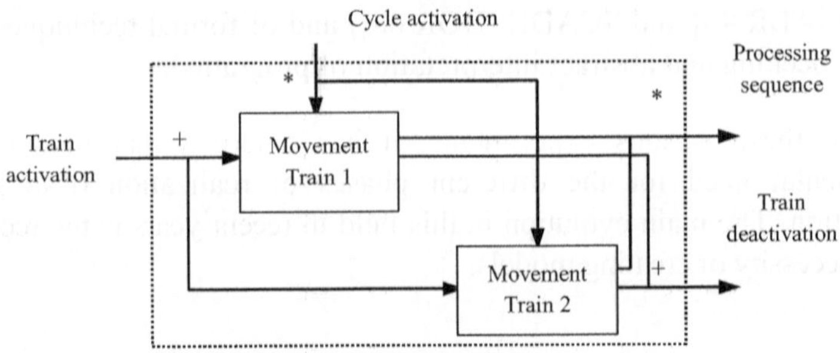

Figure 8.1. *Example of a static model introducing different types of communication*

A model is generally made up of two mutually complementary parts:

– a static model (Figure 8.1) showing the entities that constitute the system, and the different states that can be associated with those entities;

– a dynamic model (Figure 8.2) illustrating the possible changes of state.

Modeling is often based on a graphic representation of the system, which is described in the form of a tree-like structure with distinct and autonomous entities which communicate with one another, and with the environment of the system. The earliest models were based on functional analyses of the system (see the work done on SADT[1] [LIS 90]).

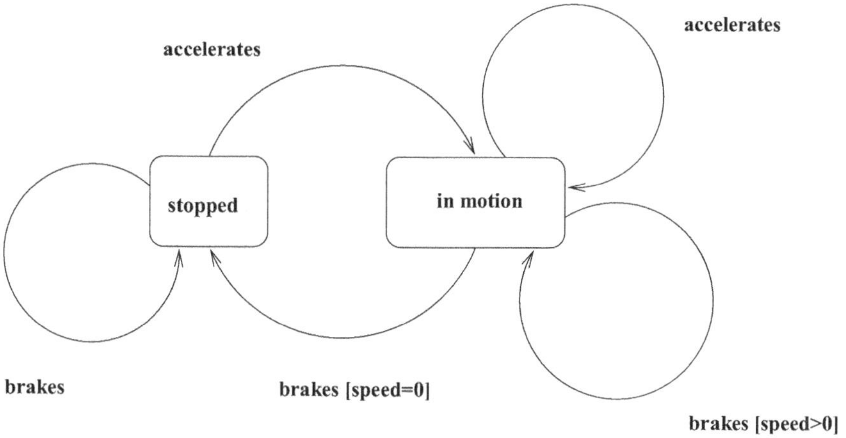

Figure 8.2. *Example of a dynamic model introducing different types of communication*

In summary, overall, modeling consists of converting a physical situation into a symbolic model in a more or less abstract language – be it iconic (graphical symbols: tables, plotted trends, diagrams, etc.) or logical/mathematic (functions, relations, etc.).

8.2.2. *Different types of modeling*

Requirements modeling can be performed at different levels and for different objectives. Thus, each level of requirements has a corresponding modeling level. We can use models of objectives or of usage to model the user requirements, functional models to model the system requirements, and finally performance models to model the architectural requirements.

1 The acronym SADT stands for *Structured Analysis and Design Technique*. The method was developed by Softech in the United States. SADT is an analysis method by successive levels of an approach describing any set of objects or data.

We have identified two phases: the analysis phase and realization phase. In the analysis phase, the objective is to identify the requirement and define it more precisely. In order to do so, we create a so-called needs model, which allows the different users to express their needs and to reason about those needs.

8.2.3. *Model*

The realization of a model M is a means to understand and/or grasp a problem/situation. Generally, the specification phase, during which the technical specifications are absorbed, involves the creation of a model M.

A model may be more or less close to the system under study; we speak of the degree of abstraction. The closer the model, the closer the results obtained will be to those which will be observed on the final system.

Another characteristic of models relates to whether or not the support language has set semantic rules.

Technique/measure	SSIL0	SSIL1	SSIL2	SSIL3	SSIL4
1. Data modeling	R	R	R	HR	HR
2. Data Flow diagrams	–	R	R	HR	HR
3. Control flow diagram	R	R	R	HR	HR
4. Finite-state machines or State Transition diagrams	–	HR	HR	HR	HR
…					
7. Formal methods	R	R	R	HR	HR
…					
11. Sequence diagrams	R	HR	HR	HR	HR

Table 8.1. *Table A.17 from the new version of CENELEC 50128:2011*

The presence of a set of semantic rules means that we can apply reasoning techniques to ensure the correctness of the results obtained.

The new version of CENELEC 50128 ([CEN 11a], Table A.3) recommends that the architecture of the software application be based on a structured method (SADT, etc.). However, it is possible to implement modeling which is based on the techniques shown in Table A.17 of the standard (see Table 8.1 here).

From these various experiments, what emerges is that modeling is an essential need for the different phases of realization of a software application. The main evolution in this field in recent years is the recognition of the necessity of creating models. Figure 8.3 shows an example of a SIMULINK model.

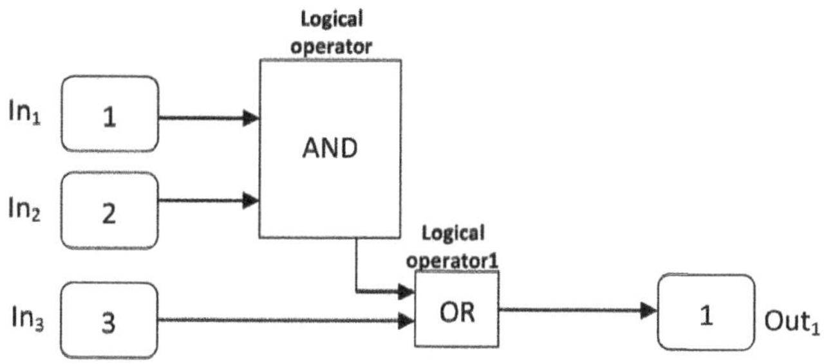

Figure 8.3. *Example of Simulink*

8.3. Use of formal techniques and formal methods

It is interesting to note that the definitions given in Appendices B and C (Part 7 of IEC 61508) use, as their points of reference, articles and/or books dating from between 1970 and 1998. It seems it is now necessary to update these to reflect the situation today.

8.3.1. *Definitions*

8.3.1.1. *Semi-formal method*

According to IEC 61508 ([IEC 08], B.2.3), a semi-formal method offers a way of developing a description of a system at a given stage of development (specification, architecture and/or design). The description can be analyzed or animated in order to check whether the system being modeled satisfies the requirements (real and/or specified). The standard notes that state diagrams (finite-state machines) and temporal Petri nets are two examples of semi-formal methods.

	SSIL0	SSIL1	SSIL2	SSIL3	SSIL4
Formal methods, including CCS, CSP, HOL, LOTOS, OBJ, VDM, Z and B, for example	–	R	R	HR	HR
Semi-formal methods	R	R	R	HR	HR
Structured methodology including, for instance, JSD, MASCOT, SADT, SDL, SSADM and YOURDON	R	HR	HR	HR	HR

Table 8.2. *Table A.2 from CENELEC 50128:2001*

The 2001 version of CENELEC 50128 introduced the concept of semi-formal methods in its Table A.2, which made reference to Table A.18.

	SSIL0	SSIL1	SSIL2	SSIL3	SSIL4
Logic/function diagram blocks	R	R	R	HR	HR
Sequence diagrams	R	R	R	HR	HR
Data flow diagrams	R	R	R	R	R
Finite-state machines/state transition diagrams	–	R	R	HR	HR
Temporal Petri nets	–	R	R	HR	HR
Decision/truth tables	R	R	R	HR	HR

Table 8.3. *CENELEC 50128:2001 – Table A.18*

In the new version of CENELEC 50128, semi-formal methods disappear as techniques. This is due to the ambiguity in determining where the boundary between what was "formal" and what was "semi-formal" lay, and certain industrialists considered that the existence of a modeling tool was sufficient to call for the use of a semi-formal method.

8.3.1.2. *Formal method*

Section B.30 of the IEC 61508 standard [IEC 08] indicates that a formal method (e.g. HOL, CSP, LOTOS, CCS, temporal logic, VDM[2] [JON 90] and

2 In the IEC 61508 standard, the references to VDM include VDM++ [DUR 92], which is an object-oriented, real-time extension of VDM. For further information, see: www.vdmportal.org/.

Z[3] [SPI 89]) enables us to construct an unambiguous and coherent description of a system at a given stage of development (specification, architecture and/or design).

The description comes in mathematical form, and can be subjected to mathematical analysis. This mathematical analysis may be tooled.

Generally, a formal method offers a notation, a technique to formulate a description in that notation and a process of verification to ensure the correctness of the requirements.

Note that IEC 61508 indicates that it is possible to perform transformations to "a logical circuit design[4]".

Petri nets and state machines (mentioned in the context of semi-formal methods) may be considered to be formal methods, depending on the degree of conformity of their usages to a rigorous mathematical base.

8.3.1.3. *Structured method*

The goal of structured methods (CENELEC 50128 – B.60) is to promote quality in software development by devoting more attention to the earlier phases in the lifecycle. This kind of method employs precise and intuitive notations (generally with computer assistance) to produce and document the requirements and the characteristics of the product's implementation.

Structured methods are tools for reflection. They are based on a logical order of thinking – an analysis of the system which takes account of the environment, the production of the system documentation, the breakdown of the data, that of the functions and of the elements needing to be verified, and must impose a certain degree of intellectual servitude (a simple, intuitive and pragmatic method). Examples of structured methods include SADT [LIS 90], JSD, CORE, the real-time Yourdon method and MASCOT.

Structured specification is a technique aiming to demonstrate the simplest visible relations between the partial requirements included in the functional specification. The analysis is refined step by step, until we obtain small,

3 Note that in IEC 61508, the B-method [ABR 96] is considered to be a method associated with Z.

4 These are the terms used in IEC 61508.

clear partial requirements. We then obtain a hierarchical structure of partial requirements which will enable us to specify the complete requirements. The method emphasizes the interfaces between the requirements, and helps to prevent interfacial faults.

8.3.1.4. *Computer-aided specification tools*

In IEC 61508, in section B.2.4, it is stated that the use of computer-aided specification tools enables us to obtain a specification in the form of a database, which can be automatically inspected with a view to evaluating coherence and completeness. The tool may allow for animation. It can be applied during the different phases of specification, architecture and/or design. For the design phase, its use is recommended provided that appropriate tools are available, but it will be necessary to demonstrate the correctness of the tool (feedback on experience and/or independent verification of results).

8.3.2. *UML*

Figure 7.7 shows an example of a UML model[5] [OMG 11, ROQ 07] of a railway system implementing different models: use cases, sequence diagram, state/transition diagram and class diagram.

It must be remembered that UML is merely a notation, and that in order to make it into a method, it is necessary to put in place a methodology which defines the elements used, the meaning associated with those elements, the points of verification, etc. This methodology must be based on a modeling guide and on a training set appropriate for each domain in question.

Hence, the use of the notation UML does raise certain issues [BOU 08a, OKA 07]. How can we use a notation which has no semantic rules? How can we evaluate an application based on UML notation? And so on. A number of publications set out to provide answers to these questions, such as [OOT 04] and [MOT 05], for instance.

The notation UML [OMG 11, ROQ 07] is not, at present, recognized as a structure and/or semi-formal method in the set of standards, although many wish to implement it either partially or completely.

---

5 For further information, see the OMG's Website: www.omg.org/.

In [BOU 07b, BOU 07c, BOU 08a, BOU 09a] and Chapter 19 of [RAM 09], we showed how UML notation can be used to create models of critical systems.

In the projects RT3-TUCS, ANR-SAFECODE and ANR-RNTL-MEMVATEX, we studied different possible ways in which to introduce UML as a means of modeling a critical system – for example, see [BOU 05, BOU 08a, IDA 07a, IDA 07b, OKA 07, RAS 08].

In recent years, a number of projects have been able to propose a formalization of the UML notation by way of a transformation to formal languages, such as [IDA 06, IDA 09, LED 02, MAM 02, MAR 01, MAR 04]. Also of note are publications such as [MOT 05], which have put forward additional rules to verify UML models.

8.4. Brief introduction to formal methods

8.4.1. *Recap*

Formal techniques (simulation, model-checking, abstract interpretation, proof, etc.) are not a new phenomenon. Indeed, the earliest papers discussing this subject date from the 1970s (for examples, see [COU 77, DIJ 76, HOA 69]). However, the implementation of formal methods dates from the 1980s [HAL 91, JON 90, SPI 89] with uses in industry emerging in the 1990s [BEH 93, OFT 97].

In [BOW 95] and [OFT 97], we find the first feedback from industrialists about formal techniques, and in particular, feedback on the B-method [ABR 96], LUSTRE [ARA 97, HAL 91] and SAO+, which is the predecessor of SCADE[6]. Other publications, such as [HAD 06, MON 00], offer an overview of formal methods from a more scientific standpoint.

The purpose of formal methods is to eliminate the ambiguities, misunderstandings and incorrect interpretations which may arise in the description in natural language.

6 It should be noted that, to begin with, SCADE was a development environment based on LUSTRE, but that since version 6, SCADE has become a language in its own right (the code generator for version 6 does indeed use a SCADE model as input, rather than a LUSTRE code).

The umbrella term "formal method" covers:

– languages:

- which have a formally-defined vocabulary and syntax;

- whose semantic rules are defined mathematically.

– transformation and verification techniques based on mathematical proof.

Formal specification invariably yields a description of what the software must do, but not how it must do it.

It should be noted that in the context of critical applications, at least two formal methods have recognized and widely-used design environments which cover part of the process of realization of code specification, whilst integrating one or more verification processes. These two methods are the B-method [ABR 96] and the programming language LUSTRE [HAL 91, OFT 97] and the graphic version thereof, called SCADE [DOR 08]. The B-method and the SCADE environment are associated with tried-and-tested industrial tools.

The advantage of formal methods lies in the fact that verifications such as proofs of properties (by means of activities of proof, exhaustive simulation [BAI 08], etc.) can be implemented as early as possible in the development cycle. The underlying idea is to create a program which is safe because of its very construction.

8.4.2. *Usage in the railway domain*

8.4.2.1. *From Z to the B-method*

In the context of SACEM[7] [GEO 90], the RATP (the group in charge of Paris transport) conducted a proof with Hoare logic [HOA 69] to demonstrate that the requirements had been taken into account (for further information, consult [GUI 90]). The Hoare proof enables us, on the basis of a program P and a set of preconditions C, to reveal all of the post-conditions.

The Hoare proof conducted for SACEM was able to make explicit a certain number of properties of the code, but it was not possible to form the link

7 *Système d'Aide à la Conduite, à l'Exploitation et à la Maintenance.*

with the safety-related requirements (such as the requirement of non-collision, for example).

In light of this situation, it was decided to construct a formal model in Z notation [SPI 89]. This formal model enabled the developers to break down the properties and form a link between the requirements and the code. Thus, twenty or so major anomalies were detected by the team of experts in charge of re-specifying the system in Z.

In the wake of the difficulties encountered during the creation of SACEM (manual processing, introduction of errors, complexity of the properties obtained, difficulty in conducting the proofs, problem of traceability, etc.), the company in charge of the development of the SAET-METEOR [LEC 96, MAT 98] was asked to use a tooled formal process integrating the formal proof. In that context, the B-method was chosen.

For that reason, in the railway domain in France, the use of formal methods, and particularly the use of the B-method, is increasingly commonplace in the development of critical systems. The software in these safety systems (railway signaling, automatic train operation) must conform to very strict criteria in terms of quality, reliability and robustness.

The B-method [ABR 96] was developed by Jean-Raymond Abrial. It is a model-oriented formal method like Z and VDM, but which facilitates an incremental development, from the specification to the code, by way of the concept of refinement [MOR 90], in a unique formalism: the language of abstract machines. At each step in the development in B, obligations of proofs are generated in order to ensure the validity of the refinement and the consistency of the abstract machine. Thus, the B-method is based on the concept of proof.

More recent projects such as CTDC, KVS, SAET-METEOR [BEH 93, BEH 96, BOU 06], the VAL at CdG Airport and Line 1 of the Paris Metro use the B-method throughout the entire development process (from the specifications to the final code – see Figure 8.4).

It should be noted that the B-method has an industrial environment. The B-workshop is marketed by CLEARSY. This tool covers the whole of the development cycle proposed by the B-method (specification, refinement, code generation and proof).

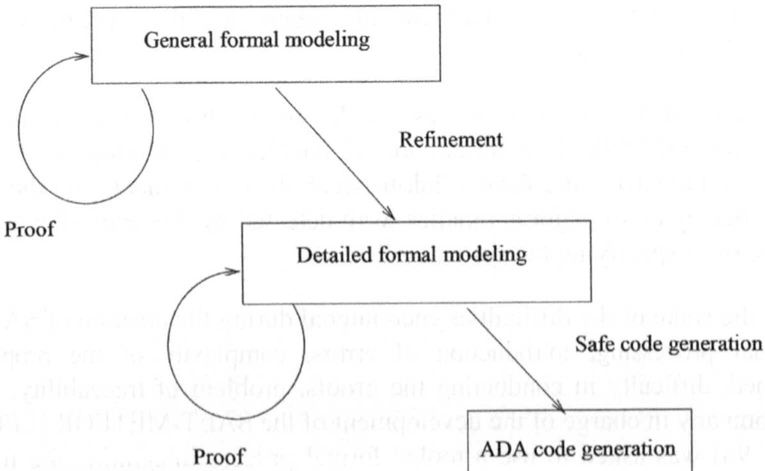

Figure 8.4. *B development cycle*

```
MACHINE
        HelloWorld
OPERATIONS
Hello =
        skip
END
```

```
IMPLEMENTATION
        HelloWorld_n
REFINES
        HelloWorld
IMPORTS
        BASIC_IO
OPERATIONS
        Hello =
        STRING_WRITE("Hello World")
END
```

Figure 8.5. *The "Hello, World!" program written in B*

[BOU 14c] is devoted to the B-method, and presents various examples of its implementation.

8.4.2.2. *From LUSTRE to SCADE*

In light of the observation that conventional computer languages were inadequate for creating automated, real-time applications, the language LUSTRE was developed at the VERIMAG[8] Laboratory in the 1980s [BEN 03, HAL 91, HAL 05].

8 For further information, see: www.verimag.fr.

LUSTRE then evolved into SCADE[9], by means of various phases and approximations (SAGA with Merlin Gérin, SAO[10] with AIRBUS). It should be noted that SAO had two points in its favor:

– code generation on the basis of automated models helped to decrease the number of errors;

– the graphic language, close to the traditions to which aeronautical engineers are accustomed, facilitates communication within AIRBUS, thus speeding up the development and helping to capitalize on the existing *savoir-faire* within the company. Hence, SAO is one of the factors in the success of the A320.

[HAL 05] offers a more detailed presentation of the history of LUSTRE and SCADE, and of the differences which exist.

The SCADE environment implements an extension of the (textual) language LUSTRE. This extension preserves the properties of LUSTRE and facilitates modeling based on the concepts of functional block diagrams and data flow diagrams. LUSTRE belongs to the category of "formal methods", in the sense that it is based on a syntax and a precise set of semantic rules. LUSTRE is a textual language.

The SCADE environment allows us to model an application in graphic form by means of components (boxes, state/transition graphs, etc.), which are combined on a "plank" (see Figure 8.6).

Safety Critical Application Development Environment (SCADE) is a language and a development environment which is used in field such as avionics (AIRBUS, Eurocopter), the nuclear sector (Schneider Electric, CEA), the railway domain (Ansaldo STS, AREVA, THALES, RATP) and has recently begun being used in the automobile industry as well.

9 SCADE is distributed by Esterel-Technologies – see: www.esterel-technologies.com.

10 SAO (*Spécification Assisté par Ordinateur*, or Computer-Aided Specification) is a "domestic" formalism developed by AIRBUS that recycles the ideas of analog block diagrams, and is therefore able to offer a specification tool and also a code generator.

SCADE is intended to be compatible with the recommendations of the various industrial standards (DO-178, IEC 61508, EN 50128 and IEC 60880).

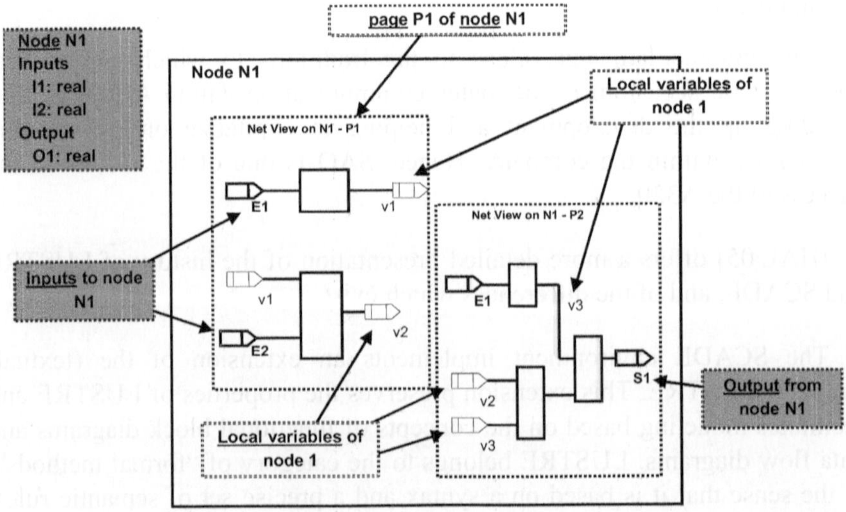

Figure 8.6. *Example of a plank*

Figure 8.7 takes the tools from the SCADE environment and places them into the V-lifecycle. Whether the SCADE environment is used for the specification or design phases, we always begin with an initial document entitled "specification" here. This document may be either a textual specification or a formal specification (SART), and may be at system level or software level, etc.

At the global level, it is interesting to put in place a model that describes the software's functional architecture. The system being modeled is then broken down as a function of the requirements, corresponding to SCADE models. This breakdown enables us to:

– distribute the global specification into various SCADE models (one model per requirement function);

– independently develop each SCADE model;

– produce the corresponding interface document.

Figure 8.8 offers a detailed view of the models used to manage SCADE nodes, and the relations which govern them.

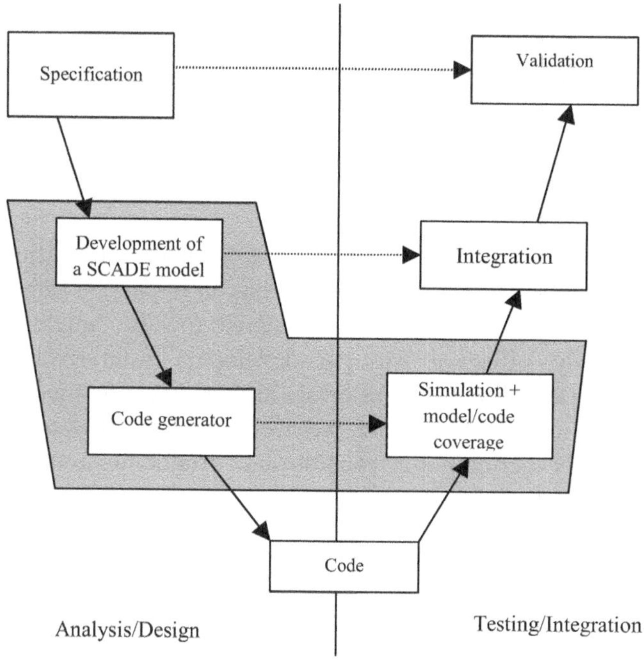

Figure 8.7. *SCADE as a design environment*

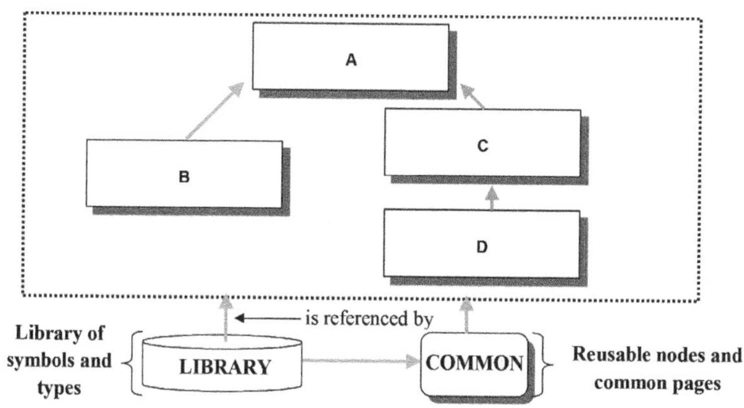

Figure 8.8. *Architecture of SCADE models*

8.4.3. *Summary*

In the collection [BOU 11a, BOU 12a, BOU 12b, BOU 14c], we presented examples of an industrial environment that can be used to implement formal methods. Although today, there are numerous environments and a huge number of applications, it must be noted that there has been a change – indeed, we have switched from "timid" experimentation with the use of formal techniques to their use on an industrial scale, as we shall explain later on.

However, this change does create a number of difficulties. The development of a software application that has an impact on safety is a rather difficult task which is very much orientated toward "quality control[11]". Indeed, the only effective way to develop a safety-related software application is to avoid introducing errors and/or detect as many such errors as possible, and in order to do that, we need quality control techniques. QC involves carefully defining the process and producing a very significant amount of documentation.

The introduction of formal methods into the process leads us from a quality-control approach to a "model-centered" approach, which may give the impression that the documentation is no longer of use, and renders the SSIL for the software application dependent of the SSIL of the tools employed.

8.4.3.1. *Model-checking*

It should be noted that formal techniques were initially used by industrialists to check that a specification, architecture, design principles and/or code respected certain properties.

As Figure 8.9 illustrates, previously, the work of checking entailed:

– identifying the properties: an analysis of the input elements (specification, architecture, realization principles, code, etc.) would be performed with the aim of identifying the properties. The properties can be classified into two families: safety properties and rapidity properties;

11 "Quality Control" in the sense of implementation of a process in line with a QC standard – e.g. ISO 9001:2008 [ISO 08].

– modeling: having chosen a technology (technique + toolkit), it is then necessary to create a model M which can be interpreted by the tools and enables is to implement the chosen verification technique (model-checking, proof, simulation, etc.);

– analyzing: the third phase consists of performing the verification itself.

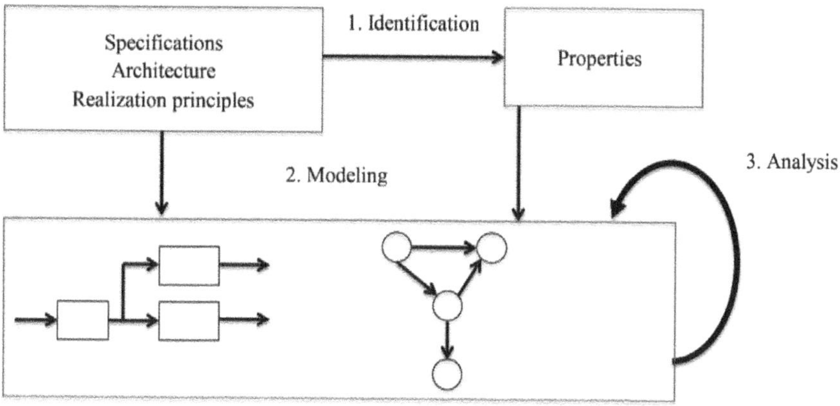

Figure 8.9. *Identification, modeling and analysis*

For example, Chapter 2 of [BOU 12a] shows how the approach illustrated in Figure 8.9 was implemented by the RATP[12] to validate the specifications of the safety functions in the SAET-METEOR[13].

8.4.3.2. *Establishment of formal methods*

Two types of approaches have been presented in the different chapters:

– the first type of approaches consists, on the basis of a specification, of realizing a formal model (see Figure 8.10) and performing verifications on that model;

– the second type consists of performing formal analyses (see Figure 8.11) on a code realized in the conventional manner (in C, ADA or C++, for example), on the basis of a specification (see Figure 8.4).

12 For further information, see: www.ratp.fr/.

13 The system [MAT 98] which, since October 1998, has been being used on Line 14 of the Paris Metro, is called SAET-METEOR (which stands for *Système d'Automatisation de l'Exploitation des Trains – METro Est Ouest Rapide*).

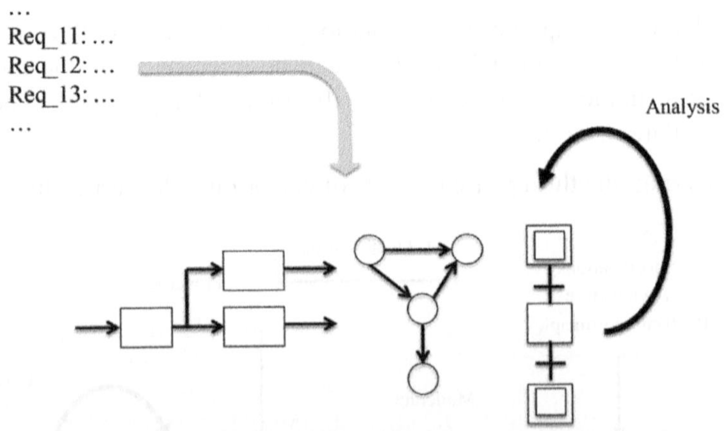

Figure 8.10. *Realization of a model*

In [BOU 11a], we presented many examples of the implementation of a static analyzer belonging to the "abstract interpretation" family (see Chapter 3 of [BOU 11a] and [COU 77, COU 00]). Thus, we presented examples of the use of FramaC, Polyspace, Astrée and CodePeer.

In order to integrate formal methods, the process of realization needs to evolve, to include the modeling phase, as shown by Figure 8.12.

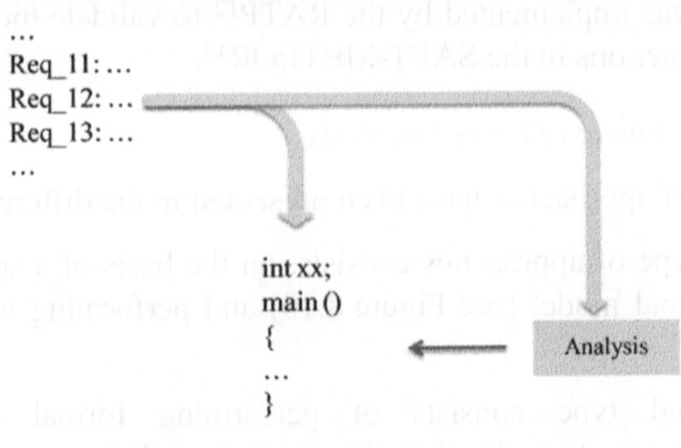

Figure 8.11. *Formal analysis of a code*

Normally (Figure 8.12(a)), on the basis of the design documents (specification, architecture, design, etc.) and the industrial procedures, the code must be produced. The inclusion of a modeling phase (Figure 8.12(b)) increases the importance of the concept of a tool and therefore the need to qualify the tools on the basis of impacts on the software application.

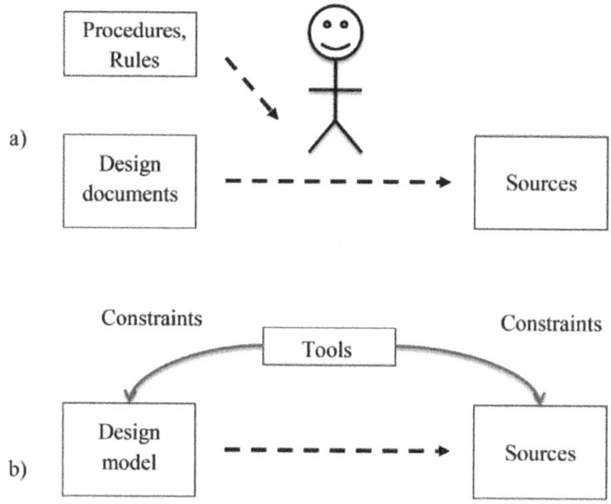

Figure 8.12. *Formal analysis of a code*

8.5. Implementation of formal methods

8.5.1. *Conventional processes*

The realization of a software application is subdivided into steps (specification, design, coding, tests, etc.). We speak of the lifecycle. The lifecycle is necessary to describe the dependencies and sequencing between the activities.

This lifecycle enables us to identify phases, and for each phase, we can identify the inputs, outputs, the activities needing to be performed, the people involved and the means of verification to be used.

Figure 7.1 shows a typical V-lifecycle. The V cycle is thus divided into two phases – the descendant phase and the ascendant phase[14]. There are a number of different test phases: unit tests[15] (focused on the lowest-level

14 It should be noted that there is a horizontal correspondence (dotted-line arrow) between the activities of specification, design and testing.

15 Unit tests or module tests or component tests.

components), integration tests (focused on the software interfaces and/or hardware interfaces) and functional tests (sometimes called validation tests) which are designed to show that the product conforms to its specification. The activities of the ascendant phase (execution of the UTs/ITs and FTs) are prepared during the descendant phase.

The operation/maintenance phase, for its part, pertains to the operational life of the product and management of any future evolutions. The maintenance of the software application remains the trickiest activity. Indeed, following any change, we must maintain a level of safety whilst staying in control of the cost of the evolution and minimizing the impact on a system in service.

8.5.2. *Process including formal methods*

The process of realization of a software application generally involves different phases requiring the application to be executed. As Figure 8.13 shows, the concept of execution is therefore essential to demonstrate that the software application functions properly.

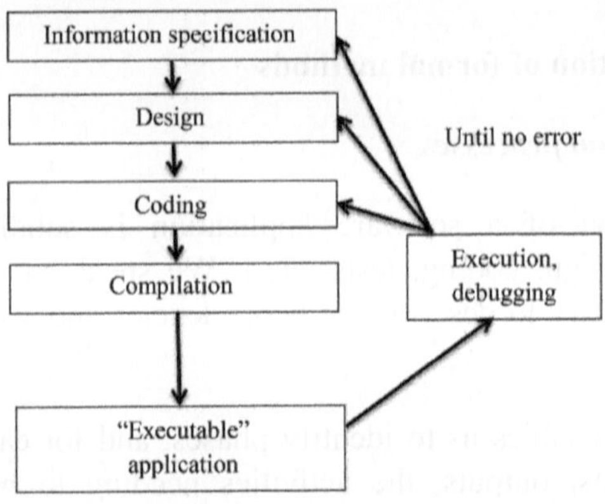

Figure 8.13. *Typical process*

In the context of development based on the concept of modeling (Figure 8.9), the process is based not on the execution of the application to demonstrate that the application works properly, but rather on the process of realization. Then, the software application is considered correct after the process that has been implemented. The correctness is therefore intrinsic to the realization of the product.

Figure 8.14 shows an example where the realization of a model is performed at the specification stage, and in this case the model is an addition to an informal specification which may contain the requirements. However, the model may also be introduced at the design level (Figure 8.15).

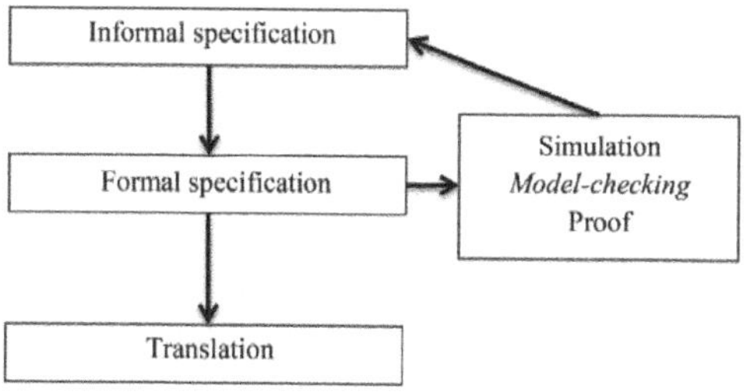

Figure 8.14. *Formal process beginning at the specification stage*

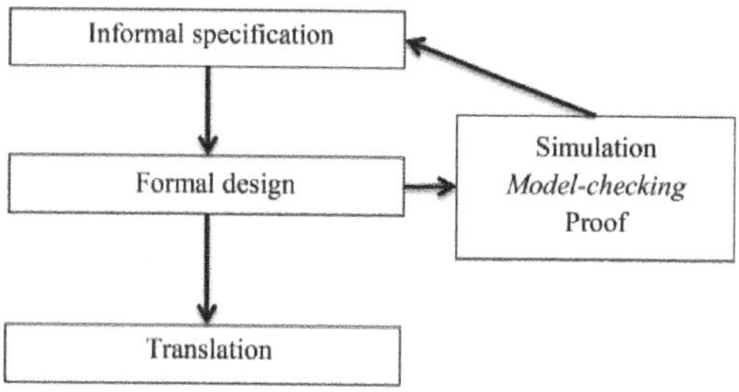

Figure 8.15. *Formal process beginning at the design stage*

In Figure 8.15, the specification is indicated as being non-formal, but it is possible to have an initial model (formal or at least structured), the goal of which is to structure the requirements and help to identify any incoherent or incomplete points. This model can then, during the design phase, be associated with a formal design model.

In the context of the development of SAET-METEOR (see Chapter 3 of [BOU 12a]), we have shown how the B-method [ABR 96] has been used to realize the design of safety-related software. The process used for SAET-METEOR was therefore one similar to Figure 8.15.

In the field of railway applications, the software specifications are generally associated with a structured model based on SADT [LIS 90] and/or SART.

It should be noted that in the development of the software of a new CBTC[16], an industrial actor in the railway domain is in the process of creating a software application where the specification is modeled with SART and the design combines SCADE with the B-method.

8.5.3. *Issues*

The introduction of formal techniques and formal methods in the realization of software applications causes the apparition of new processes which are no longer in phase with the V-lifecycle recommended by the standards currently in force. The realization cycle using formal methods shifts the verification activities to the downward phase and concentrates all the information in the model (see Figure 8.16).

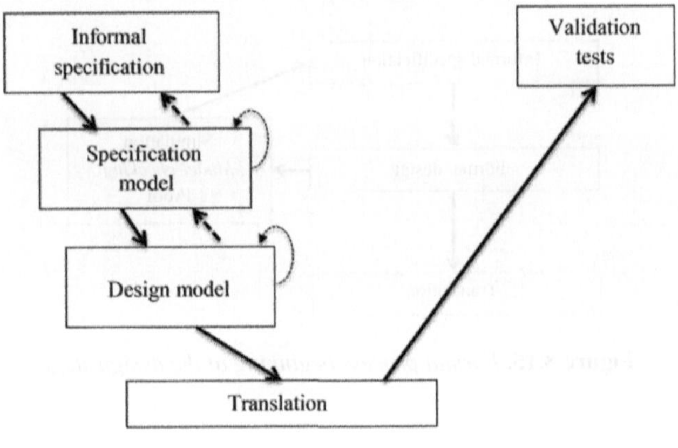

Figure 8.16. *Process involving models*

The concentration of the information in the model leads us to say that this is not a "model-oriented" approach, but rather a "model-centered" one, which is different. Indeed, with a "model-centered" approach, the model becomes the backbone for all the activities, or it comes to contain all the

16 *Communication-Based Train Control*: this is a system controlling the operation, driving and safety on trains and Metros. CBTC is a system comprising devices installed aboard trains and fixed devices that communicate between themselves (generally by radio). CBTC is subject to a standard [IEE 04].

activities (in Figure 8.16, the dashed arrows correspond the verification activities).

The quality-control standards (ISO 9001:2008 [ISO 08], for example) recommend that we identify the processes, activities and documents needing to be produced. The industrial standards such as CENELEC EN 50128 [CEN 01a, CEN 11a] identify a minimal list of documents to produce.

In a model-centered process, the UTs and ITs can be replaced by simulations and/or proof of property. However, these verification elements are then an integral part of the model, as are the elements such as:

– the calculation of the metrics;

– the verification that the model is coherent;

– the verification that the coding rules have been respected;

– etc.

This type of “model-centered” approach is interesting, but it does away with numerous documentary elements which are then felt to be “unnecessary”, because they can be produced on demand. In that, we have the potential for “non-conformity” to “objective”-oriented standards such as IEC 61508 [ISO 08], ISO 26262 [ISO 11], la CENELEC EN 50128 [CEN 01a, CEN 11a], etc.

The strength of the “model-centered” approach is that it helps the professionals involved to come together, as we showed in [BOU 99, BOU 06]. Indeed, the various teams (software and hardware, design, V&V and safety) can pool information relating to their activities, which helps to foster a better understanding of the difficulties. Generally, the team in charge of safety, which uses its own functional models and its own methods, may have a slightly different vision to that of the design team, and those differences may have a very significant impact in terms of costs and time. The use of a shared model enables the safety-related recommendations and requirements to be communicated to everyone involved, as soon as possible, and it is also possible to link those elements to specific parts of the model.

The "model-centered" approach will[17] cause four problems:

– the use of models at all levels and of all types. The difficulty relating to this point lies in the absence of semantic rules for the modeling tools (UML notation, SIMULINK and/or MATLAB model, etc.);

– the lack of documents means it will not be possible to rework the design in case of obsolescence of the tools;

– the tools become the central point in the process, so it is necessary to demonstrate that the tools are usable for the SIL applicable to the software application under development. This qualification must be applied to the code generators as well as to the verification tools;

– the complexity of the techniques employed (for example, see abstract interpretation tools [BOU 11a]) requires a high level of skill of the people in charge of the activities.

8.6. Maintenance of the software application

The maintenance of the software application remains the trickiest activity. Indeed, following any change, we must maintain a level of safety whilst staying in control of the cost of the evolution and minimizing the impact on a system in service.

The maintenance of a software application comes up against a difficulty: the lifespan of that software application. Indeed, for the railway domain, the projected lifespan is 40-50 years, in aeronautics it is 40 years, in the nuclear domain 50 years, but in the automotive industry, it is only around 15 years.

In light of these lifespans, we need to take measures to ensure the maintenance of service and the maintenance of the software application. These measures must take account of the type of machine used for the development, the tools employed, the documentation needing to be produced to perform the maintenance and repeat if need be.

Système d'Aide à la Conduite, à l'Exploitation et à la Maintenance (SACEM) is a system that was commissioned in 1986 and has regularly been updated since. The choice of a language such as ADA [ISO 95, BAR 14] or

17 We use the future tense here, because at present there is no model-centered application that has been in operation long enough to cause problems, but it could come to pass.

C [ISO 99] helps to arm against the problems of obsolescence of the machines, the operating systems and the tools.

But what will become of formal development environments such as the B-tools and/or SCADE? Will it be possible to transfer the model from one tool to another? Will it be possible to redevelop a tool?

8.7. Conclusion

Formal techniques and formal methods are now successfully used, industrially, on projects of varying sizes. The tools associated therefore have attained a certain level of maturity which means they are able to deal with the complexity of such applications. Note that the complexity of industrial applications very frequently has an impact on processing time[18].

The use of formal techniques and formal methods has an impact on the process that is conducted, and it is therefore necessary to construct a new referential in line with the quality standards in force. The construction of that referential must take account of the fact that the model can replace documents which it used to be essential to produce, but also that we must take account of the lifetime of the system and the associated maintenance objectives. Indeed, if the tool is no longer maintained and the formalism is proprietary, it may be difficult, or even impossible, to update the application in the absence of the model.

Another difficulty with model-centered approaches is that they make the SSIL of the software application dependent on the SSIL of the tools used. If it is difficult to demonstrate that a C compiler is SSIL4, it will be even more difficult to show that a prover is SSIL4. In the context of compilers, it was possible to implement strategies based on redundancy and diversity. However, for specific tools such as provers, model-checking tools and/or tools with abstract interpretation, it is difficult to find two tools with similar efficacy in the same domain[19]. Therefore, we need to establish tool qualification files.

---

18 In the context of SAET-METEOR, it took over a week in 1998, with the tool Polyspace (see Chapter 8 of [BOU 11a]) to analyze the 100,000 lines of code in ADA, whereas the same task would now take 1-2 hours.

19 It must be remembered that for many of the tools used today, the algorithms are not public property, and are even subject to copyright.

The various standards (DO 178, ISO 26262, IEC 61508, CENELEC EN 50128) include, or will soon include, the concept of a “qualification file”. The qualification of a tool depends on its impact on the final product.

The situation is not reduced to a single vision. The main advantage to the B-method [ABR 96] and SCADE 6 [DOR 08] is that they offer an approach which leads from the software specification to the coding phase (passing through the architecture and design steps along the way). The code generator in the B workshop is, for certain versions, considered to have been qualified through use, and the code generator in SCADE 6 includes a certificate.

As regards formal techniques and formal methods, the question of qualification of the tools is absolutely crucial, because the complexity of the technologies used (provers, model-checking, etc.), the confidentiality aspect (licensed algorithms, etc.), the innovation aspect (new technology, relatively few users, etc.) and the maturity aspect (product stemming from research, product created under an “open” license, etc.) mean it is not easy to build confidence in a tool.

9

Tool Qualification

9.1. Introduction

In this chapter, we shall present the process of tool qualification, as defined in the 2011 version of CENELEC 50128. The previous version focused on the qualification of compilers, but the new version takes account of the set of tools used for the realization of a software application. Thus, the idea of tool qualification is presented, and examples are used to explicitly express the need.

The realization of a piece of software is based on the use of software for the tasks of development, compilation, verification and validation. The standards that are applicable for the realization of critical software define a process to manage the introduction of faults into the programs, but that process can only deliver on its predefined objectives if the tools themselves also respect specific objectives.

The B version of the standard DO-178 introduced the concept of tool qualification, and the concept was taken up again in the 2008 version of IEC 61508, and therefore in the standards derived from it, such as ISO 26262 and the 2011 version of CENELEC 50128.

The goal of this chapter is to present the issue, the concepts associated therewith and the expectations.

9.2. Concept of qualification

9.2.1. *Issue*

Initially, the development of a software application consisted of producing a code. In that case, in order to obtain an executable file, we have to use a compiler. If the process used to obtain the code conforms to an SSIL objective, the fact of the software application conforming to its SSIL objective is directly linked to the compiler. We need to demonstrate that the compiler can be used for an SSIL objective.

More generally, this issue can be generalized to the set of tools used for the design, compilation, verification and validation of the software application, but also the tools used in the production of the data for the software application.

As we have explained, it is essential to demonstrate that a tool is usable in the realization of a software application with a predetermined SSIL objective. This activity is called the qualification of a tool.

DEFINITION 9.1 (Qualification).– *The qualification of a tool is a process whereby it is demonstrated that a tool is usable for the realization of a software application with a definite SSIL.*

In the coming sections, we shall give a more detailed presentation of exactly what qualification is, explicitly discussing the inputs, outputs and activities.

9.2.2. *CENELEC 50128:2001*

In Article 10.4.7 (section on design), CENELEC 50128:2001 stated that we must choose an appropriate set of tools (design methods, language, compiler) for the SSIL, for the whole of the software's lifecycle.

More specifically, Article 10.4.9 introduced a stipulation regarding the selection of the language to be used – the compiler had to have one of the following articles:

– a validation certificate;

– an assessment report;

– a redundant process.

With certain languages, such as Ada (e.g. see [ANS 83, ISO 95, BAR 14]), that have a standard that describes the language, it is possible to establish a certification process. For example, as regards the certification of Ada compilers, the existence of a standard and of a fairly detailed set of semantic rules for the Ada language has made it possible to define a process for certifying a compiler. This process has been carried out on various compilers. It is based on a series of tests known as Ada Conformity Assessment Test Suite (ACATS) – see [ISO 99]. For further information, consult [ADA 01]. This approach, although it is applicable, is still confidential.

The second possibility is to draw up an evaluation report detailing the appropriateness of using the given compiler in the particular context. This approach is what we call “qualification” of the compiler.

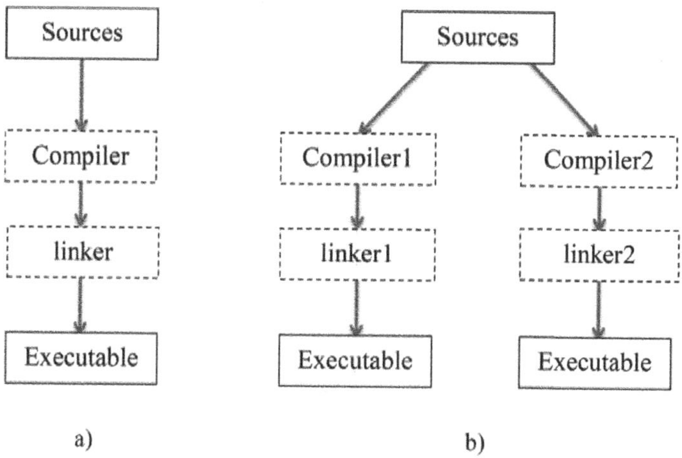

Figure 9.1. *Two examples of a compilation sequence*

Figure 9.1 shows two examples of the executable-generation sequence. (a) represents a linear sequence where a single executable is produced from the sources, unlike (b) which shows the production of two executables. Approach (b) enables us, on the basis of a redundant process, to prove that the compiler does not adversely impact the executable.

For the third case, the standard was more specific, as it prescribed “a redundant process based on a signature check which enables us to detect errors in translation”. This approach was directly inherited from the *PSC*

(*Processeur Sécuritaire Codé* – Coded Safety Processor) used in the railway domain in France (see [BOU 10a], Chapter 2).

The PSC was used in the development of SACEM, and has been deployed on many similar systems. The PSC is based on a redundant process which is able to produce an executable file and a series of codes on a hardware architecture capable of executing the application, whilst also performing a check on the codes predetermined offline. Although it is interesting, this type of redundancy is very specific and is not widely used.

Technique/Measure	SSIL0	SSIL1; SSIL2	SSIL3; SSIL4
…			
11. Translator validated	R	HR	HR
12. Translator tried, tested and trusted	HR	HR	HR
…			

Table 9.1. *Table A.4 from CENELEC EN 50128:2001*

As Table 9.1 shows, in the area of design, we tend to focus on the translator (in the sense of a compiler) with a link the section B.7, which made the switch from a validated translator to a certified compiler, with a generalization to all tools.

Note that row 12 of the table, regarding a tried-and-trusted translator, is linked to section B.65, which identifies the concept of a tried-and-trusted compiler. This concept allows us to attach value to the experience the developers have acquired with a given version of a compiler. This experiment needs to be formalized, and the projects (nature, complexity, SSIL, etc.) and the known anomalies must be identified. The consequence of this approach is to limit the introduction of new languages and/or new compilers.

We can conclude from this that in CENELEC 50128:2001, there was a stipulation that the compilers must be qualified, but no real indication of exactly what was required.

9.2.3. *DO-178*

9.2.3.1. *Version B*

Version B of the DO-178 standard discusses the need to put tool qualification in place (see section 12.2). DO-178 necessitates the qualification of the tools if the process identified in the standard is impacted (removal or reduction of activities, automation, etc.) by the use of a tool.

DO-178:B introduces two categories of tools:

– software development tools: the outputs are integrated into the onboard software application;

– software checking tools: these tools do not introduce any new errors, but they may be incapable of detecting those errors which are present.

DO-178:B indicates that the qualification is valid for a given use on a specific system, so there is no capitalization.

Thus, the qualification process consists of:

– identifying the type of tools: development tools (CC1) or verification tools (CC2);

– identifying the level of the tool: the tool's software level must be the same as that for the onboard software application, unless we can justify a reduction of that level. The reduction of the level is connected to the type of the activity upon which it impacts, to the importance of that activity, to the probability of being able to detect an anomaly by another means, etc.;

– implementing a quality-assurance process similar to that which needs to be established for an onboard application;

– having a specification of the tool;

– checking that the tool conforms to its specification – section 12.2.1-d of DO-178 describes the necessary activities;

– validating the tool's behavior. It is possible to monitor the tool over a certain length of time in order to confirm that the outputs are correct;

– managing the faults: any and all faults present in the tool must be logged, analyzed and corrected;

– establishing a configuration management process similar to that which needs to be implemented for an onboard application.

During the qualification process, described below, detailed records must be kept.

For each tool, a qualification plan needs to be drawn up in order to identify the configuration of the tool, the desired level (CC1 or CC2), the required level Design Assurance Level (DAL), a description of the architecture, an identification of the activities to be performed and an identification of the data needing to be produced.

Note that the qualification must be approved by the certifying authority.

9.2.3.2. *Version C*

In version C of the DO-178 standard [RTA 11a], several documents were introduced, including one devoted to tool qualification, which is known as DO-330. DO-330 retains the CC1 and CC2 categories, and introduces a qualification objective – the Tool Qualification Level (TQL) – on a scale of 1 to 5.

DO-330 proposes a radical change in philosophy in relation to version B of DO-178, and also in relation to the other standards (IEC 61508, ISO 26262, 50128). Indeed, it is suggested that developers establish a tool realization process which is not the DO-178 standard, but rather a specific process, stripped back in comparison to the stipulations of the industry-wide standard. This approach is innovative and more realistic.

9.2.4. *IEC 61508*

In its 2008 version, IEC 61508 [IEC 08] identifies three classes of tools: T1, T2 and T3. These three classes are defined as follows:

– T1 generates no output that is able to contribute, directly or indirectly, to the executable code (including the data) for the safety-related system;

– T2 takes care of the testing or verification of the design or the executable code, when errors in the tool itself may prevent it from detecting faults, but cannot directly create any errors in the executable software;

– T3 generates outputs which are able to contribute, directly or indirectly, to the executable code for the safety-related system.

Based on these classes, IEC 61508:2008 identifies the need to put a qualification process in place.

9.2.5. *ISO 26262*

In its 2011 version, the ISO 26262 standard also identifies three classes for tools, referred to as the TCL (Tool Confidence Level). TCL2 and TCL3 are the two classes which require the establishment of a set of actions to be selected on the basis of the ASIL (Automotive SIL) of the software.

Tool qualification can be performed as the combination of four activities:

– feedback on experience;

– evaluation of the tool realization process;

– establishment of a validation of the tool;

– realization of the tool in accordance with ISO 26262.

9.3. CENELEC 50128:2011

9.3.1. *Introduction*

The 2011 version of CENELEC EN 50128 formally introduces the need to qualify the tools (see section 6.7 of the standard). As with IEC 61508 and ISO 26262, three classes of tools are introduced: T1, T2 and T3. The three classes are defined as follows:

– T1 generates no output that is able to contribute, directly or indirectly, to the executable code (including the data) for the software;

– T2 facilitates the testing or verification of the design or the executable code, when errors in the tool may prevent it from detecting faults, but will not directly create any errors in the executable software;

– T3 generates outputs which are able to contribute, directly or indirectly, to the executable code (including the data) for the safety-related system.

The definition of T2 is the same as it is in IEC 61508:2008 [IEC 08]. For T1, we focus on the software, and for T3, we introduce further data.

Ultimately, class T1 contains tools which do not have an impact on the verification and which have no impact on the final executable – examples include:

– editing tools (text editors, model editors, etc.) which do not have the ability to generate elements to be integrated into the code;

– additional tools such as configuration management tools.

Class T2 is given over to tools where an error could skew the results of the verification or validation. The T2 category contains tools for checking the coding rules, measuring the metrics, for static code analysis, test management and execution, coverage measuring tools, etc.

Class T3 is devoted to tools whose failure could have an impact on the final executable. This class includes compilers, code generators, data preparation tools, etc.

9.3.2. *Qualification file*

Section 6.7 of CENELEC 50128:2011, for each class, defines a set of recommendations which identifies the contents of the qualification file.

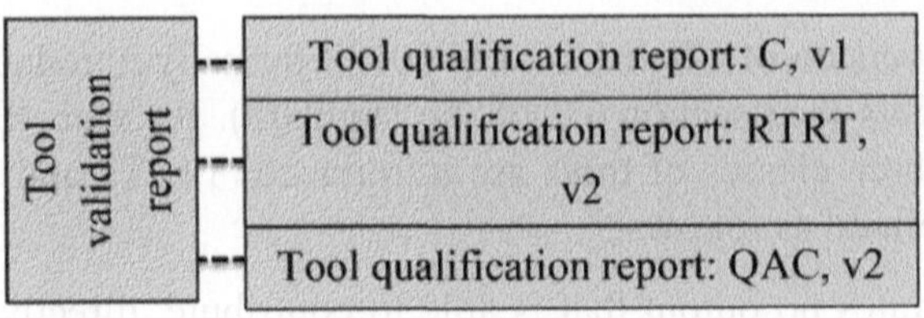

Figure 9.2. *Tool validation file versus qualification file*

CENELEC 50128:2011 proposes the preparation of an overall document known as the tool "validation report". Thus, this report must cover all of the tools involved in the project. However, in the context of an industrial project, it is preferable to compile standalone files, and for those files to be managed by the tools rather than by the project team. The tool validation report (TVR) at the project level is therefore merely a framework (see Figure 9.2) which can be used to reference the qualification files (with reference and version).

9.3.3. *Qualification process*

CENELEC 50128:2011 identifies 12 requirements (numbered from 6.7.4.1 to 6.7.4.12) concerning tool qualification. Requirement 6.7.4.12 is linked to Table 9.2, which has been corrected in relation to the published version. We have enriched the table with steps that enable us to classify the requirements needing to be implemented. These steps will be listed and explained in the next section.

Table 9.2 shows a progression in the efforts needing to be mobilized, but it is incomplete because it does not take account of the SSIL. IEC 62279:2014 [IEC 14] is the image of CENELEC 50128. The version of IEC 62279:2014 being finalized at the time of writing gives us Table 9.3.

On the basis of Table 9.2, we can identify a methodology for the qualification of the tools which is divided into five activities:

– identification: we must identify the tool and the version used;

– justification: it is necessary to identify and justify the class Tx of the tool;

– proof of conformity: the tool must have a specification, and it must be demonstrated that the tool conforms to that specification;

– appropriateness for the needs: we need to demonstrate that the tool is in line with the overall methodology being used for the software development;

– version management: we must manage the configuration of the qualified version, and stay abreast of known anomalies.

For each item 6.7.4.x, IEC 62279:2014 [IEC 14] introduces additional specifications, but these do not change the essential points of the requirements set out in CENELEC 50128:2011.

Class of tools	Sections applicable	Step
T1	6.7.4.1	Identification
T2	6.7.4.1	Identification
	6.7.4.2	Justification
	6.7.4.3	Specification
	6.7.4.10, 6.7.4.11	Version management
T3	6.7.4.1	Identification
	6.7.4.2	Justification
	6.7.4.3	Specification
	((6.7.4.4 and 6.7.4.5) or 6.7.4.6)	Conformity proof
	(6.7.4.7 or 6.7.4.8) and 6.7.4.9	Requirement fulfillment
	6.7.4.10, 6.7.4.11	Version management

Table 9.2. *Table 1 from CENELEC EN 50128:2001*

Class of tools	Methodology	SSIL
T1	None	-
T2 T3	Demonstration of successful use in similar environments	1-2
	Support process implemented in addition to the tool	
	Execution of a library of test cases, which can be used to demonstrate that the tool works properly	
	Respect of all the requirements in CENELEC 50128 for the development of the tool	3-4
	Justification of the existence of a process of implementation of the tool which demonstrates the SSIL	
	Execution of a recognized test case library to demonstrate that the tool works properly	
	Execution of a process using several different tools with a process of comparison	

Table 9.3. *Table 3 from the draft version of IEC 62279*

9.3.4. *Implementation of the qualification process*

9.3.4.1. *Identification*

This first activity is fairly simple, in the sense that it consists of identifying the tool which we wish to use (name, reference, functionalities, etc.) and explaining precisely why that tool is necessary.

When choosing the tool, we must take account of its insertion into the process of realization of the software application. Indeed, we need to explicitly illustrate the appropriateness of the use of this tool, and its conformity with the SSIL objectives and the functionalities that are or can be used.

Even in the 2001 version of CENELEC 50128, there was already a demand to show that the design process was appropriate for the SSIL objectives. In the 2011 version, we are asked to go further – to think about the whole process, and therefore about the cooperation between tools, compatibility of files and formats, the goal being to minimize the number of errors (caused by the tools, the exchanges and the manipulations).

There are various different types of tools: specification tools, modeling tools, code generators, compilers, chargers, test environments, configuration management tools, etc.

Figure 9.3 illustrates an example of a development process, showing the editing tools, the executable-generation tools, the V&V tools and the configuration management tools.

One of the difficulties in the implementation of the tools lies in the number of functions that they provide. Hence, we must be able to define the functional perimeter that is to be qualified (see Table 9.4). It is generally difficult to obtain a complete specification of each tool, and the effort needed for validation is therefore proportional to the number of functions being used.

9.3.4.2. *Justification of class*

On the basis of the information gleaned during the identification phase, it is necessary to identify the impact of a failure of the tool, and therefore the class Tx associated with each tool.

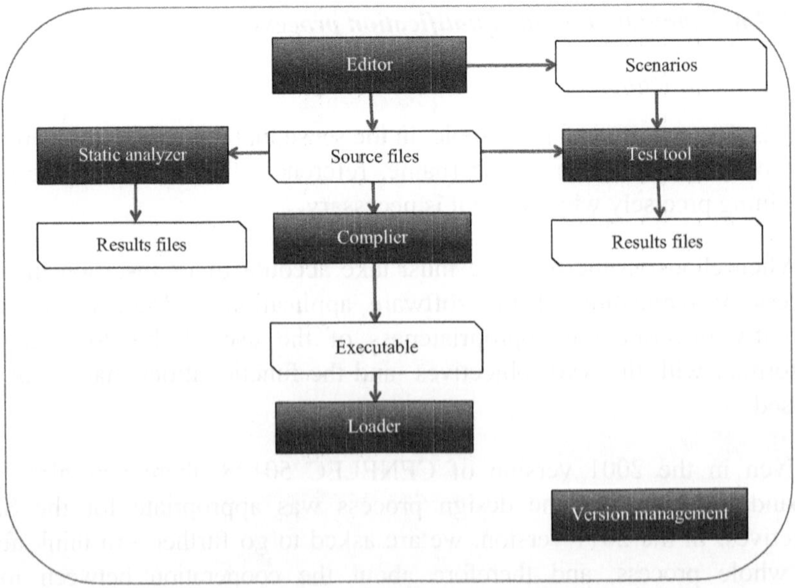

Figure 9.3. *Tools in the process*

<table>
<tr><th>Function</th><th>Level 1</th><th>Client</th><th>Target</th></tr>
<tr><td rowspan="3">Configuring</td><td>Defining a configuration</td><td>X</td><td></td></tr>
<tr><td>Sending a configuration</td><td>X</td><td>X</td></tr>
<tr><td>Simulating devices</td><td></td><td>X</td></tr>
<tr><td>Simulating</td><td>Reading/writing</td><td>X</td><td>X</td></tr>
<tr><td rowspan="3">Editing a scenario</td><td>Adding a scenario</td><td>X</td><td></td></tr>
<tr><td>Modifying a scenario</td><td>X</td><td></td></tr>
<tr><td>Removing a scenario</td><td>X</td><td></td></tr>
</table>

Table 9.4. *Example of identification of the functions of a tool*

9.3.4.2.1. Choice of Tx

For each tool, on the basis of its role in the process of realization of the software application, we need to identify the associated Tx. This step is tricky because certain tools may play different roles in different projects.

For example, in project A, a spreadsheet can be used to edit the requirements (with number, attributes and description), and in this context it would be T1. For project B, the spreadsheet can be used as a tool for visualization and analysis of the test results, or as a generator of test reports; in that context, the same tool would be classed as T2. For project C, however, that same tool can be used as a means of generation (using macros) of data files in ASCII format, in which case it would be classed as T3. This example with a spreadsheet illustrates the difficulty inherent in the task of qualification, and the fact that we need to analyze the use of the tool in the process of realization of the particular software application.

9.3.4.2.2. Failure analysis

The second step in this phase involves carrying out a simplified dysfunctional analysis, because we are seeking not to study the tool's safety, but rather to identify the impact of a failure of that tool on the overall process.

Table 9.5 offers an example of dysfunctional analysis of the configured and simulated services which were identified in Table 9.4.

Table 9.6 shows a more complete typical plan to perform a dysfunctional analysis at tool level, but this is merely a suggestion.

9.3.4.2.3. Compiler

In this section, we present a second example of dysfunctional analysis, focusing on a compiler. Out of the possible failures in a compilation sequence, the potentially-hazardous situation identified is as follows:

> Production of erroneous assembly codes in relation to the high-level instructions (e.g. in the language C) to which the compilation pertains.

In Table 9.7, we offer a simplified analysis of the potential faults of a compiler, and the hazardous situation is therefore associated with the concept of "degraded function". For this failure, various methods of risk reduction are put forward: qualification, double sequence, verification of a code by a coded safety processor (PSC – see [BOU 09], Chapter 2).

9.3.4.3. *Specification*

As indicated earlier on, it is necessary to define the function boundary of tools classed as T2 and T3 (see Table 9.4, for example) and to have a specification and/or a user manual that documents the functions being used.

The specification elements must identify the usage constraints: type of machine, type of OS, memory size, usable options, etc.

ID	Function	Sub-function	Failure mode	Effect on performance of a test	Detection	Risk reduction measure	Remark
1	Configure	Configuring of a simulation	Establishment of an incorrect configuration	Scenario unexecutable	Detectable		
				Scenario executed	Detectable		Incorrect configuration
				Scenario executed	*Undetectable*	Verification of configuration before and after execution of the scenarios	T2
2	Simulate	Execution of a simulation	Incorrect interpretation of commands	Scenario unexecutable	Detectable		
		Playing of a scenario		Scenario executed	*Undetectable*	Establishment of tests	T2

Table 9.5. *Example of fault analysis*

ID	Function	Sub-function	Phase of life	Failure mode	Effect	Detection	Safety	Risk reduction measure	Remark
1	Fx	SFx	Creation Modification Launch All	No function Degraded function Untimely function Partial function Unexecutable function Delayed function	NA Partial element …	Yes No	nS - S	Tests Verification …	T1 T2 T3 …

Table 9.6. *Typical dysfunctional analysis*

9.3.4.4. *Proof of conformity*

As indicated in Table 9.2, proof of conformity of the tool is necessary if it is classed as T3.

9.3.4.4.1. Approach

CENELEC EN 50128:2011 introduces three requirements, which must be read as follows: ((6.7.4.4 and 6.7.4.5) or 6.7.4.6), contrary to what is indicated in Table 9.7 of the standard.

ID	Function	Failure mode	Effect	Detection	Safety	Risk reduction measure	Remark
1		No function	No executable	Yes	nS	–	No execution
		Partial function	No executable	Yes	nS	–	No execution
2	Compiler	Degraded function	Executable	No	S	Validation process	–
						Qualification	T3
						Double compilation sequence	–
						Checking of a code against runtime errors	Coded safety processor type approach
						...	
3		Untimely function	No executable	Yes	nS		Generation is impossible with the control of the operator
4		Delayed function	Executable	No	ns	–	No constraint on compile time

Table 9.7. *Example of analysis of a compiler*

As Figure 9.4 shows, Clause 6.7.4.4 identifies the general approach to demonstrate the tool's conformity, which is based on five approaches: proof of correct operation (list of previous uses, certificates, etc.), validation, implementation of a means of execution verification (e.g. double compilation and comparison, code checking to detect compilation errors, etc.), use of a process to detect errors in the tools, and any other methods – manual or otherwise (for example, one could manually read through a piece of code that has been generated).

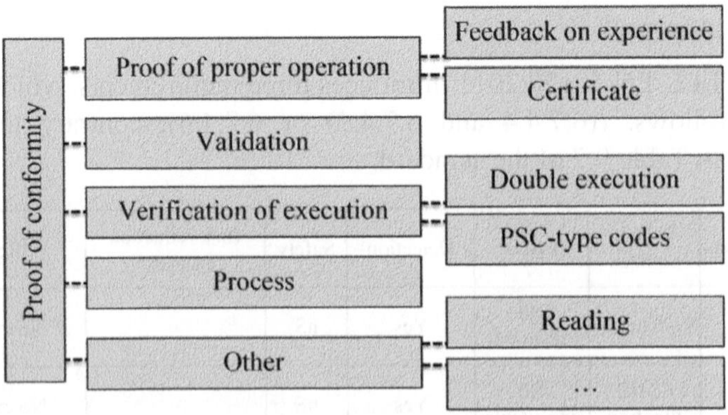

Figure 9.4. *Different approaches to proof of conformity*

9.3.4.4.2. Proof of proper operation

Proof of proper operation can be performed by using a product that has a certificate or by demonstrating that we have past experience from which we have learnt.

9.3.4.4.3. Certified product

Indubitably, the existence of a certificate is the easiest way to verify a tool's quality, but we still need to verify that the certificate is applicable to the use at hand, and that the exported constraints have been taken into account. Verification that the certificate is applicable is done by way of the activity of cross-acceptance (Application Guide for CENELEC EN 50129 – Part 1 [CEN 07]), which needs to be performed by an independent evaluator.

"Cross-acceptance" is a phase which involves checking that the normative framework, the hypotheses and the constraints exported from an evaluation report and/or from a certificate are compatible with the planned use for the project.

9.3.4.4.4. Documented feedback

Documented feedback involves demonstrating that the tool works properly by formally documenting previous experience. This approach enables us to benefit from previous uses of the tool and to prepare for the future. This way of working is in line with the concept of "high level of confidence demonstrated through use" that is introduced in Table E.9 of CENELEC 50129:2003 (see Table 9.8).

Techniques/measures	SIL2 - SIL2	SIL3 - SIL4
High level of confidence demonstrated through use Optional when prior proof is not available	R: 10,000 hours of operating time and at least a year of experience in operation with the equipment in place	R: 1 million hours of operating time, at least two years of experience in operation with the different equipment in place, including safety analyses, with detailed documentation of the minor changes made during the time in operation

Table 9.8. *Table E.9 from CENELEC EN 50129:2003 [CEN 03]*

The difficulty with documented feedback stems from the (very, and often too, rapid) evolution of the tools; it is rather difficult to ensure that a tool's behavior remains the same as it evolves. From one version to the next, it is possible to correct and inadvertently add faults which may or may not have an impact on the tool's behavior and the usable services. Thus, it is important to clearly identify the services that are or could be used, and to have a process for approval of any new versions in place (section 6.7.4.11 of CENELEC 50128).

Thus, for documented feedback, we need to put in place a continuous process. This process must recommend the use of a tool that has already been qualified, define a procedure for approval of a new version and recommend that the documented feedback be updated at the end of each project.

The documented feedback may be documented in the form of a table, as illustrated by Table 9.9.

Tools	Version	Function	Project name	Number of lines	SSIL	Date	Remark
B tools	3.04	Code generation	METEOR	100,000	SSIL4	Oct. 98	First use

Table 9.9. *Example of documented feedback*

In the railway domain, as regards compilers, there is currently a certain amount of experience that can be documented as feedback; these practices have meant that tools are only modified in case of absolute necessity.

9.3.4.4.5. Validation

The validation of a tool is identified in clause 6.7.4.5 of CENELEC 50128:2011. This clause recommends that we define a validation strategy, which should focus on the list of usable services.

As is true of the validation of a software application, the validation of a tool is based on a set of tests which must completely cover the specification (or usage manual) and the list of usable services. The resources used must be identified so as to be able to repeat the tests identically. The test scenarios and results must be logged so they can be reviewed and audited. We must then formulate and formally record a judgment as to whether the tool's services do indeed work properly. In case of anomaly, that version of the tool may be rejected, or a new restrictive perimeter may be defined.

9.3.4.4.6. Other

Documented feedback and validation are the two favored approaches, but they are not always practicable. Indeed, with complex and innovating tools – e.g. automatic proofing tools and abstract interpretation tools – we are unlikely to be able to have any feedback, and/or it will be difficult to establish a validation (particularly when the algorithms involved are subject to copyright).

In this situation, it is necessary to carry out a detailed analysis of the possible faults, and imagine whether it is possible to manage these faults:

– by a process of realization (combination of tools and activities);

– by verification of the execution (e.g. a double compilation and comparison of the runtime sessions);

– by additional activities such as manual checks.

Clause 6.7.4.6 in the standard enables us to formalize a justification, which must be documented and will be subject to the evaluator's approval.

The first difficulty with this approach is linked to re-use; the argument may depend very heavily on the project. The second difficulty stems from the fact that this approach is subject to discussion, and there is no guarantee of being able to reach an agreement with the evaluator as to the acceptance of the method.

9.4. Fitness for purpose

9.4.1. *Design method*

CENELEC 50128:2011 recommends that the software design method be compatible with the features of the application.

There are different types of paradigm that can be used to realize a design and/or a code, such as the object-oriented approach (UML, OMT, etc.), the functional approach or the imperative approach. We must demonstrate that the chosen approach is compatible with the application being realized (onboard software, etc.). The idea of compatibility is linked to the fact that an inappropriate approach could render the realization and comprehension of the model more difficult, and therefore in turn make it more difficult to maintain the software.

The design method and the programming language should enable errors to be detected as early as possible – e.g. at compile time. For this reason, C is less appropriate than ADA, as ADA [ADA 83, ADA 01, ISO 95] is capable, at compile time, of detecting problems of typing and/or faulty construction, and offers typing control at run time.

The design can be based on code generation tools that are able to produce a partial code (signature of services, etc.) or a complete one. The use of automated code generation must be explicitly justified.

The compilation must use tools that have been evaluated as compatible with the objectives – i.e. T3 tools.

9.4.2. *In case of incompatibility*

If the design method and/or the language used are not compatible with the type of application in the process of realization, it will then be necessary to identify the weak points of the approach and come up with measures to compensate for them.

For example, the main difficulty with C lies in the weak typing the language involves. Typing is essentially the manipulation of types of bases, such as *int* (or similarly *word*, etc.). Thus, we must compensate for the lack of typing linked to a functional – e.g. x is an element of 1…10 – by additional checks which can be introduced in the form of defensive programming (*if* ($x \geq 1$ & $x \leq 10$) at the start of each function. The effect of this defensive programming, though, is to make the code more complex and therefore render maintenance more difficult.

9.4.3. *Code generation*

In the context of the use of so-called "formal" methods, or at least with model-oriented approaches, it is possible to generate the code automatically on the basis of a model. CENELEC 50128:2011 requires that when we choose to use automated code generation, we give an explicit justification as to why the tool is appropriate for the development of a safety-related software application.

If a model is used, a number of verification activities may be performed on that model. By putting the activity of automated code generation in place, it is possible to obtain a code which is similar to that of the model at input. Unless we can demonstrate that the generated code is the exact image of the model (e.g. by using a certified code generator), it is necessary to perform all the same checks as if the code had been written manually. If the code is shown to be the exact image of the model, it will then be possible to justify foregoing checks on the generated code.

Note that code generation may be employed to produce the code for the generic application and/or the code relating to the parameter data for the generic software application.

9.5. Version management

For each tool, we must be aware of the usable version(s) and the usage constraints associated therewith (see Figure 9.5).

9.5.1. *Identification of versions*

We need to identify each version that is qualified. This identification must be as accurate as possible. It is important to clearly identify any and all "patches" which may have been applied in order to obtain the qualified version.

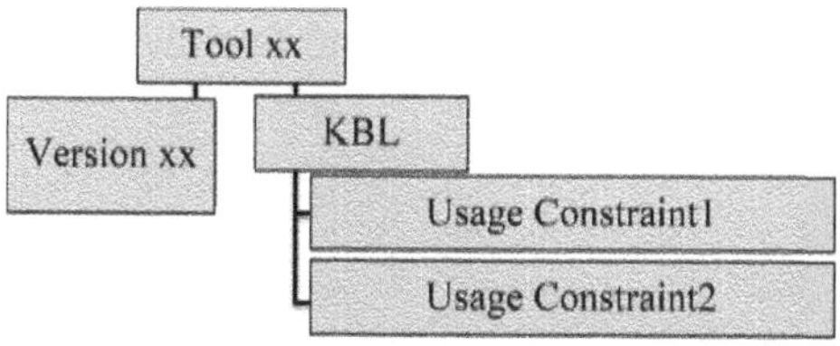

Figure 9.5. *Version management*

9.5.2. *Bug/defect analysis*

For each version, there is a “Known Bug List” (KBL). It is necessary to analyze that KBL in order to identify the constraints of use of the tool. Those usage constraints may pertain to functions which are not to be used, precautions to be taken for the tool’s implementation, or limitations (e.g. the maximum number of files the tool can process in a session).

9.5.3. *Changing versions*

The process of version management involves identifying the process of acceptance of a new version. This process is based on:

– identification of the new version;

– identification of the evolutions: functional evolutions, list of options, etc.;

– demonstration that the evolutions do not affect the compatibility of the tool with the process being implemented on the project(s);

– analysis of the new KBL (resolution of faults and therefore removal of the use constraints associated therewith, finding of new faults and addition of new use constraints).

As regards the third point, it may be wise to repeat the test which served to determine the initial qualification.

9.6. Qualification process

9.6.1. *Qualification file*

The qualification process as defined by CENELEC 50128:2011 must give rise to a qualification file associated with each tool. This qualification file must be compiled not at project level, but rather at organizational level, so that it can be used on multiple projects, and thus help enhance confidence in the tools as they are employed.

9.6.2. *Ultimately*

The 2011 version of CENELEC 50128 is not yet obligatory, but it is preferable to implement it because, unlike the 2001 version, it defines a formal framework for the qualification of the tools. Indeed, it sets objectives and identifies resources that need to be used.

9.6.3. *Qualification of non-commercial tools*

The process described in section 9.3.3 is easily applicable to commercial tools, but it is more difficult to apply it to tools developed locally, or for a specific project. In general, for non-commercial tools, it is necessary to implement a realization process that conforms to CENELEC 50128:2011.

As CENELEC 50128:2011 is oriented toward onboard applications, it is not always easy to apply it completely. It is for that reason that the approach put forward by DO 330 [RTA 11b] could be of interest, although it is not directly applicable for the railway domain.

9.7. Conclusion

One of the main advances made by CENELEC 50128:2011 [CEN 11a] relates to the introduction of classes for the tools involved in the process of realization of a software application and the definition of qualification objectives for each class.

In this chapter, we have presented the need and laid out elements used to create the qualification file for each tool.

As indicated above, IEC 62279:2014 [IEC 14] gives its Table 10.3 in addition to the demands of CENELEC 50128:2011. In the preparation of the new version of CENELEC 50126 [CEN 12], section 7.9 of Part 4 raises the number of requirements from 12 to 17. The new requirements do not alter the process that is discussed in this chapter; they pertain to the need to select appropriate tools, to have tools which cooperate and are compatible with the product or the process, the need for the tools to be available, etc.

10

Maintenance and Deployment

10.1. Introduction

Section 9 of CENELEC 50128:2011 is dedicated to the maintenance and deployment of software. This section represents a significant evolution of the standard.

10.2. Requirements

10.2.1. *Fault management*

In the context of operational safety, a fault is an element which, if executed, could lead to failure. In the world of software, we speak of software faults or "bugs". One of the few things that are certain about software is that it will invariably contain an unquantifiable number of faults. Software faults are unavoidable – they are simply part of the software world.

One of the main reasons for the evolution of a program is fault correction. Figure 10.1 illustrates the fault analysis process.

The fault management cycle consists of several phases: identification of faults, analysis of their effects (impact on the safety and/or reliability of the software application), selection of anomalies to correct, analysis of those anomalies, implementation of the corrections and verification that those corrections have been successful (usually, to verify the proper implementation of the changes, we conduct a series of tests to check that no further changes have been made).

The anomalies are analyzed by way of an impact assessment (see Definition 5.3) and a non-regression analysis (Definition 5.4). In some cases, the non-regression is said to be total, and in such cases, it is necessary to repeat all of the tests from one or all of the phases. The objective of non-regression analysis is to minimize the cost of a new version.

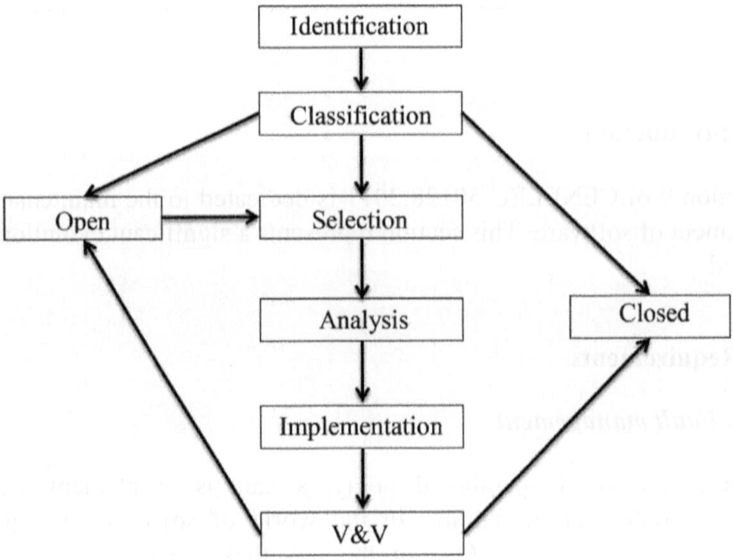

Figure 10.1. *Fault management cycle*

10.2.2. *Managing changes*

It is only possible to manage changes made to the software if that software is maintainable and testable. These two properties are identified as essential by CENELEC 50128, but they are quite difficult to implement.

One of the peculiarities of railway systems is their lifespan. The lifespan of this type of system is 40 years, with a tendency to last closer to 50 years. This raises the question of how to maintain a software application over the course of 40 years.

Figure 10.2 shows that, in general, a software evaluation needs to be repeated several times, which is why we refer to version Vx'. A delta evaluation can take account of a variety of changes (see version V2+V3).

This incremental process raises several questions:

– how do we manage the commercialization of products when we have several different certificates?

– after several delta certifications, is it necessary to perform a full certification (and if so, which are the decisive criteria for the awarding of that certification)?

– etc.

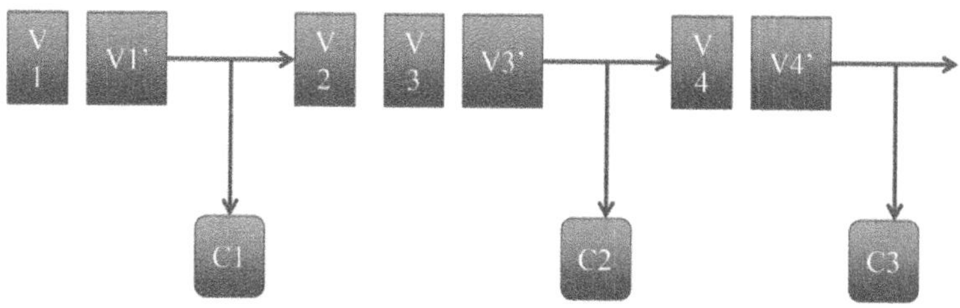

Figure 10.2. *Series of certifications*

It is easy enough to ask the above questions, but in order to find an answer, we must implement specific processes to manage the changes and the impact of those changes on the products already in use. Hence, we must have a set of principles that help prevent and/or manage incompatibilities (software/OS, software/software, hardware/software compatibility issues).

For this reason, CENELEC EN 50128:2001 has a section entitled "Software maintenance" (Chapter 16), which is very poorly understood by industrialists, who feel it is not their responsibility to manage the maintenance of the product over the course of the system's (40-50 year) lifespan, and limit themselves to correcting any software faults.

Chapter 16 of CENELEC EN 50128:2001 merely sets out the basics of evolution management; it does not specify how to establish software maintenance protocols. Furthermore, in terms of techniques, it presents only two techniques, as shown in Table A.11 (Table 10.1). For this reason, in the

2011 version of the standard, the chapter on maintenance is linked to the chapter on deployment.

Measure	SSIL0	SSIL1;SSIL2	SSIL3;SSIL4
1. Impact analysis	R	HR	M
2. Data logging and analysis	HR	HR	M

Table 10.1. *Table A.11 from CENELEC EN 50128:2001*

10.3. Deployment

10.3.1. *Issue*

CENELEC 50126 [CEN 00] recommends that we monitor the safety of the system, not just during its realization, but right up until the system is decommissioned (see Figure 2.4). The effect of this is that it is necessary to monitor and manage any modifications that are made, and to demonstrate that the SSIL is still the same after modification.

This requirement has a number of impacts on the processes employed:

– it is necessary to build integral maintenance capabilities into the system during the design;

– it is necessary to define the maintenance processes during the realization phase, and those processes must be assessed by the independent evaluator;

– any modification must be identified and authorized;

– the process of modifying a system that is actually in operation must adhere to the maintenance plans;

– the deployment of any given version must be planned and authorized;

– following a deployment, the new configuration must be logged, and the operating procedures must take account of any new exported constraints. Note that it is possible that some exported constraints from the previous version may need to be removed.

Thus, CENELEC 50128:2011 calls for software deployment management in its section 9.1.

10.3.2. *Implementation*

A new version of a software application is made available by the provision of a software version sheet (SwVS), identifying:

– the version of the software;

– the software configuration (list of applicable documents, list of sources, configuration management elements, etc.);

– the exported constraints;

– the data characterizing the interface elements: hardware version, operating system version, interfacing software versions, etc.;

– etc.

Software deployment involves replacing an existing version with a different version. This action requires a deployment process to be defined. This deployment process must be accompanied by a deployment manual (SDM).

The deployment of a new version may lead to a number of different situations:

– all goes well (which is rarely the case);

– the configuration of the equipment (train, device, system, etc.) is not the same as the one referenced in the deployment documents, and the deployment fails (improper behavior, inability to start, etc.);

– the previous version of the software is not installed properly (different files, write-protected files, files that have been moved, directories with a different name, directories not included in the tree diagram, etc.), making it difficult to install the new version. This may result in poor execution or erroneous execution (wrong library used, etc.);

– etc.

The software deployment manual must identify:

– the version of the software being deployed;

– the loading process for that version;

– the process to verify that the version has loaded correctly (verification by auto-identification of the software);

– the process to verify that the loaded version works properly in the normal environment. Thus, we must identify a number of functional tests and produce test results that will be formally recorded in the deployment log;

– the reversal process: if a problem arises that means the system cannot function properly with the new version, we must be able to “rewind” to the original configuration;

– the equipment configuration update process. Updates may be made in document form (file, train file, etc.), electronic form (database modification), etc.

The deployment of a software application thus requires a number of different documents to be produced:

– a software deployment plan (SDP), describing the deployment management method and the actions to be taken to manage this activity;

– a software deployment manual (SDM) that outline the process for a given version;

– a software deployment record (SDR) that contains the results of a given deployment.

As previously indicated, as all activity is checked, it is necessary to produce a verification report (SDVR) which shows that the productions from the activity of deployment have been verified.

10.3.3. *In reality*

CENELEC 50128:2011 was right to introduce this chapter on deployment of new versions of software, but the deployment of a new version is rarely carried out by the same industrialist that actually realized the system. Also, the developer will be able to provide the SDP, the SDM and SwVS but there is very little chance that they will be able to retrieve the records of the deployment and finalize the deployment verification report.

A manufacturer must, therefore, consider a version that has been sent to the client as having been deployed. Note that it would be wise to supplement the deployment records with a deployment log that would allow the industrialist to check all of the versions that are considered deployed.

For his/her part, the person in charge of the system's operation must implement a configuration, management process, log the elements provided by the industrialist and use a deployment process that will produce the elements requested by the client.

It will be the responsibility of the independent overall system evaluator to verify full compliance with section 9.1 of CENELEC 50128:2011.

10.4. Software maintenance

10.4.1. *Issue*

Maintenance of a software application is a real challenge. Indeed, it is not only a question of being able to alter the code, but it is necessary to ensure continuity of services offered by a software application over a relatively long period of time, using equipment which may be very variable.

The tricky part is not managing an application, but instead managing the demands of multiple customers using different version of the same software. The software deployment report mentioned in section 10.3.3 becomes very useful. When a new version is decided upon, we must take the time to check whether the modification request or fault correction is compatible with all the deployed versions, or whether we run the risk of creating a branch that would cause incompatibility issues. This decision must be carefully considered, because it could have a significant financial impact on the project.

CENELEC 50128:2011 formalizes the context of software evolutions, forming a strong link between them and the evolutions of the system (see the link with phase 13 of CENELEC 50126 [CEN 00]). The maintenance of a software application is described in section 9.2 of CENELEC 50128:2011.

10.4.2. *Implementation*

The first change is the establishment of a software maintenance plan (SMP) to describe the methodology employed to produce a new version of the software during the maintenance phase. This plan must identify the resources, methods and productions, but also the authorities who decide on a change and approve the modified software.

CENELEC 50128:2011 is not retroactive, but future developments will have to conform to it. Before modifying the software, we must decide whether the modification qualifies as major or minor, and this judgment will need to be approved by an independent evaluator.

If the modification is deemed to be minor, then section 9.2 is applicable. If it is considered major, then CENELEC 50128:2011 is applicable *in its entirety*.

Every modification must be logged. The log must contain the modification request, an analysis of the impact that change will have on the overall system, a detailed description of the modification itself, and a record of the V&V process used for the change.

A software maintenance report must be established when the software is first released, with a view to keeping track of the different versions, forming the link with all the modifications that have been made, forming the link with impacts on the system and having at our disposal a history of the software configuration.

As previously indicated, as all activity is checked, it is necessary to produce a verification report which shows that the productions from the activity of maintenance have been verified.

10.5. Product line

The management of a product line by an industrial actor will involve the establishment of variants. This is particularly true for a product that has been created in conformity with the legislation in one country and must be installed and operated in another country.

Figure 10.3 illustrates the ideal case (ideal, that is, from a development point of view, rather than from a financial one). Based on a product P characterized by a version V1 and a certificate C1, we can proceed to alter and tweak the product and create a variant which goes further in its development. In this case, the lack of interaction between the two projects will have an impact on costs (the two versions developed from the same base), and may also impact on safety; the faults detected for one project are not necessarily shared with the other project.

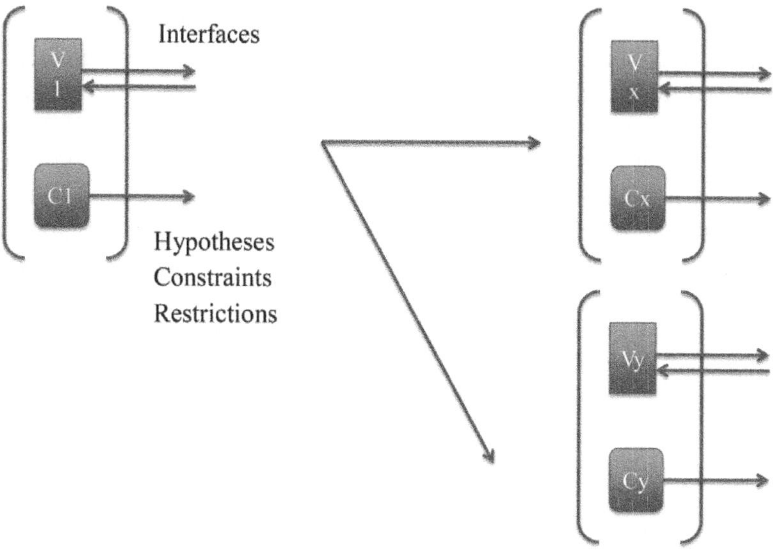

Figure 10.3. *Establishment of a variant*

As is shown by Figure 10.4, in order to remedy this sort of problem and rationalize the costs, the software application can be split into two parts: a generic part, which is shared by all the projects, and a specific part.

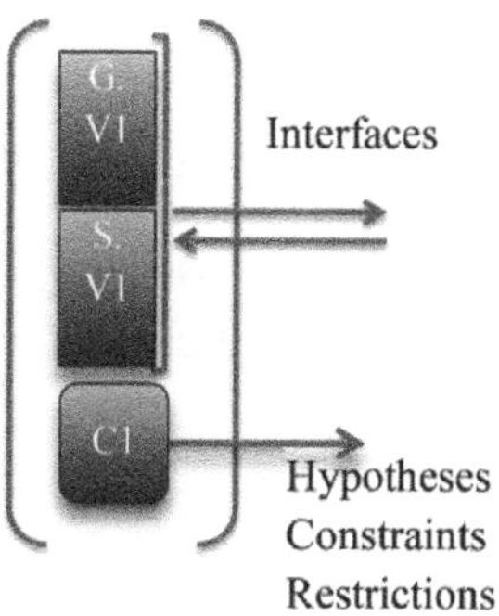

Figure 10.4. *Characterization of a specific product*

The definition of a generic application [BOU 99, BOU 00] is still a tricky point. Does the generic part need to contain all possible behaviors, or should it include as little information as possible?

If the generic part contains all possible behaviors, we find ourselves with a final software application that contains an enormous amount of dead code. In general, this code is dead code because of the software parameters. The existence of dead code is prohibited, or should at least be minimized, in the context of a critical safety application that impacts on human lives. To demonstrate that this dead code is harmless, it is necessary for the whole of the generic application to have been checked and validated. Most of the time, it is not possible to fully check (validate) a generic application including all potential behaviors.

If the specific application contains only the minimum, the important behaviors will be in the specific parts, and the workload needing to be done specifically for each project becomes rather greater. In addition, when beginning a new project, we must ensure that we use the version of the specific product which is closest to our target application.

The best way of working is to adopt a position halfway between these two extremes: to define a generic application of reasonable proportions and, as new specific application arise, offer additions to that generic application.

Regardless of the choice made, the use of parallel development for the realization of variants must identify a process of fault reporting between one variant and another. A fault that is considered dangerous for one variant may, in fact, be harmless for a different variant, but when it is discovered, it is necessary to perform the analysis.

The introduction of product lines and the introduction of a separation between the generic part (reusable, adaptable) and the specific part are not discussed in CENELEC 50128:2011. However, the standard does allow for re-use and management of development costs.

10.6. Conclusion

In this chapter, we have presented software deployment and maintenance, which are two important subjects that are beginning to emerge as points that are difficult to manage. A portion of the complexity is induced by the increase in the number of programs used in a system, which, when coupled with the number of known faults and the number of identified hardware configurations, becomes a real headache.

10.7. Appendix: documentation needing to be produced

Document title	Acronym
Deployment	
Software Deployment Plan	SDP
Software Deployment Manual	SDM
Software Version Sheet	SwVS
Software Deployment Records	SDR
Software Deployment Verification Report	SDVR
Maintenance	
Software Maintenance Plan	SMP
Software Change Records	SCR
Software Maintenance Records	SMR
Software Maintenance Verification Report	SMVR

11

Assessment and Certification

11.1. Introduction

In both the 2001 and 2011 versions of CENELEC 50128, the concept of independent assessment is explicitly introduced [CEN 01a, CEN 11a].

11.2. Evaluation

11.2.1. *Principles*

The assessment [BOU 06, BOU 07a, BOU 09b] of a product (a system or software application) entails conducting an analysis of all the components of that product, with a view to assessing the product's conformity to a given frame of reference (usually a standard, part of a standard or a set of standards), in accordance with a given method. An independent evaluation is thus achieved by an Independent Safety Assessor (ISA), who should, ideally not belong to the company having made the product.

DEFINITION 11.1 (Assessment).– *Product evaluation involves assessing the product's conformity to a specific frame of reference. There is a predefined process that must be followed for conformity analysis.*

The important thing to remember, from Definition 11.1, is that the assessment of a product involves two elements: a frame of reference and all the elements produced during the realization of the product.

Figure 11.1 shows the overlap between the development and the assessment of the software application. It also illustrates the product of the assessment process: namely the assessment report.

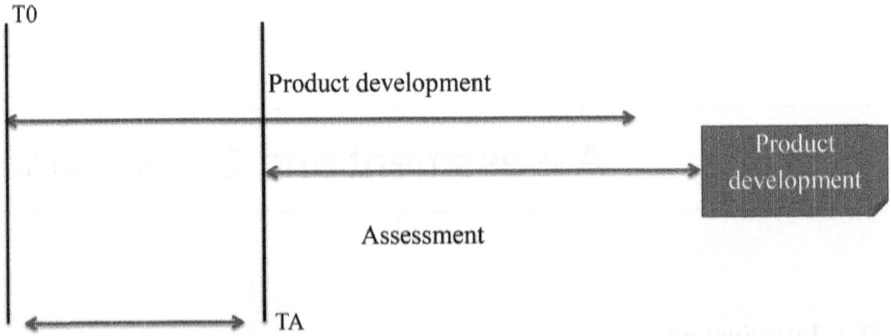

Figure 11.1. *Development and assessment*

As shown in Figure 11.1, there is a time lag between the start of the project (T0) and the start of the independent assessment (TA). If T0 = TA, the independent assessor would not have very much to do at the start of the contract. If TA were significantly later than T0 (or even, in the worst-case scenario, after the project has come to an end), the independent assessor's work may cause an overload of work for the rest of the team, or even the repeating of the entire development cycle. Ultimately, the date chosen to start the assessment (TA) must be appropriate so that the assessor has enough to work with and to allow the development team to take on board any demands for adjustment.

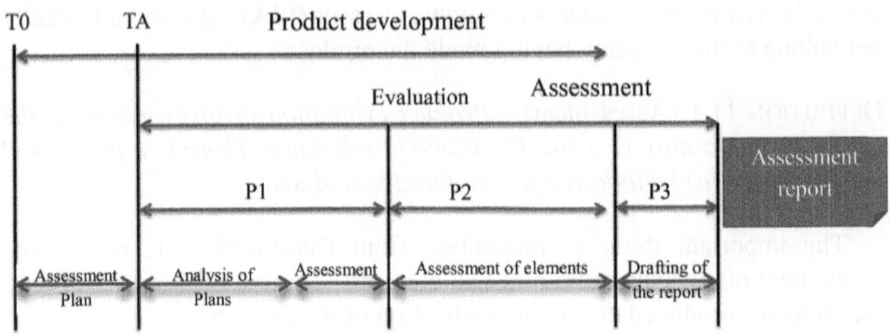

Figure 11.2. *The three phases of assessment*

Figure 11.2 shows the assessment process and the four main phases:

– drafting the assessment plan, which describes the organization, the input frame of reference, methodology and scope of the assessment;

– the audit phase, which involves analyzing the methodological plans, preparing an audit plan and performing the audit;

– the phase of assessment of all the elements produced (see Figure 11.3) as a result of the software realization process (descendant phase, ascendant phase, safety analysis, etc.);

– the final stage is to write the assessment report.

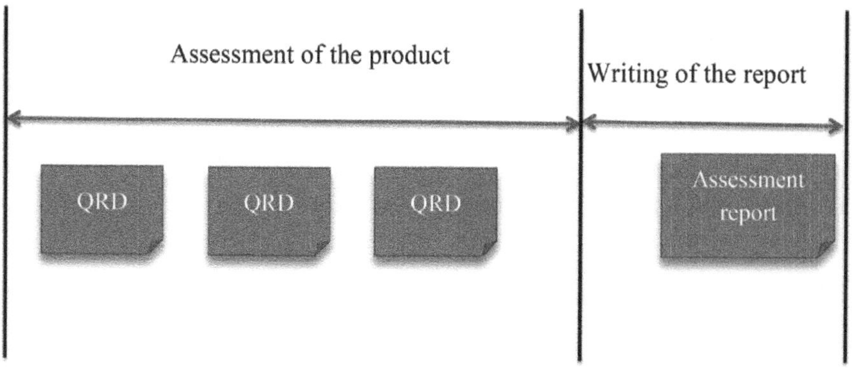

Figure 11.3. *Evaluation*

The task of assessment involves analyzing all the elements and producing question-and-remark sheets (QRSs). QRSs represent a means of conducting an exchange with the client; they are used for the formalization of questions and remarks, and of responses from the client. The client's responses may lead to proposed modifications.

The result of an assessment should state whether or not the product conforms to each of the requirements of the standard:

– if the product meets the requirement, the evaluation says it "conforms to the requirement";

– if the product does not meet the requirement, the assessment says there is a "remark";

– if the product meets all of the requirements of a standard, then the evaluation says it "conforms to the standard";

– if one or more requirements are not met, the product cannot be declared to conform to the standard.

There may be several kinds of remark, classified on the basis of the "risk incurred" if the requirement is not met.

This classification of "risk incurred" is highly subjective, and it is usual for standards to clearly state which requirements are "mandatory", which "optional" and which "recommended".

This enables the assessor to conclusively classify the remarks:

– a deviation from a "mandatory requirement" jeopardizes the product's conformity to the standard;

– a deviation from a "recommended requirement" does not, on its own, call into question the overall conformity of the product to the standard.

11.2.2. *CENELEC 50128:2011*

Section 6.4 of CENELEC 50128:2011 identifies the need to assess software applications whose SSIL is between 1 and 4. For a software application with SSIL0, if it is possible to demonstrate that it conforms to ISO 9001:2008, no assessment is necessary.

Note that Table B.8 in Appendix B of CENELEC 50128:2011, identifies the skills required of the person in charge of the evaluation. That table (reproduced below as Table 11.1) illustrates the need for recognition by an authority, which gives rise to a restriction on all assessors.

As is the case with the other activities, the assessment must be formalized in an evaluation plan (EvalP) and the results need to be formalized in an evaluation report (EvalR).

Having analyzed the whole of the product, the assessor must then identify all the points of non-conformity to the requirements of CENELEC 50128:2011 and pass judgment as to their impacts on the safety of the software. Everything must be documented in the assessment report (AsR).

It should be noted that if an application has an evaluation, the person in charge of assessment must accept that assessment after verifying that the results are applicable. This verification is performed by way of *cross-acceptance* – see the next section.

Role: assessor
Skills:
1. shall be competent in the domain/technologies where assessment is carried out
2. shall have acceptance/license from a recognized safety authority
3. shall have/strive to continually gain sufficient levels of experience in the safety principles and the application of the principles within the application domain
4. shall be competent to check that a suitable method or combination of methods in a given context have been applied
5. shall be competent in understanding the relevant safety, human resource, technical and quality management processes in fulfilling the requirements of EN 50128
6. shall be competent in assessment approaches/methodologies
7. shall have analytical thinking ability and good observation skills
8. shall be capable of combining different sources and types of evidence and synthesizing an overall view about fitness for purpose or constraints and limitations on application
9. shall have overall software understanding and perspective including understanding the application environment
10. shall be able to judge the adequacy of all development processes (like quality management, configuration management, validation and verification processes)
11. shall understand the requirements of EN 50128

Table 11.1. *Table B.8 from CENELEC 50128:2011*

11.3. Cross-acceptance

As is shown by Figure 11.4, a product is not only the description of a set of interfaces, but also a set of hypotheses, constraints and restrictions.

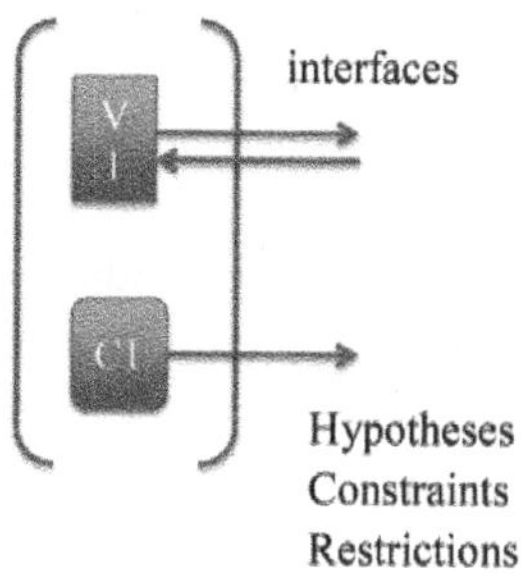

Figure 11.4. *A product*

Thus, as a general rule, the activity of design of a railway system using certified or assessed products requires us to verify the compatibility of the interfaces and of the constraints, the hypotheses and the usage restrictions. If the industrial standard used for assessment/certification is not that of the railway domain, we also need to verify that the standards are compatible.

Cross-acceptance is a phase which consists of checking that the normative context, the hypotheses and the constraints exported from an evaluation report and/or a certificate are compatible with the use intended for the project.

The application guide for Part 1 of CENELEC 50129 [CEN 07] describes the principles of cross-acceptance.

11.4. Certification

11.4.1. *Product certification*

Certification is an administrative act which expresses confidence acquired by the competent authority as to the system's aptitude to fulfill its safety-related functions.

In the railway sector, certification is tantamount to authorization to operate.

Certification is based on the "proof" of a product's safety produced for the attention of the regulatory authority by the responsible interlocutor defined in the organization document.

DEFINITION 11.2 (Certification).– *Certification consists of awarding a certificate, which is an undertaking to guarantee that the product respects a normative standard. The certification is based on the results of an assessment and the production of a certificate.*

A certificate enables us to define a boundary of usage and a boundary of responsibility. It should be noted that without a certificate, a business may be asked to demonstrate that the product can be introduced into the system that the business controls. In this case, we speak of homologation.

11.4.2. *Software certification*

It is possible to certify a product because it refers to something complete (hardware, mechanics, software, … , processes). For a certain amount of time now, certification bodies have been being asked to issue certificates for a software program.

This request makes no sense: the software on its own is not a product – it is incapable of executing itself, and its execution is dependent on the hardware that is to receive it. Nevertheless, a number of organizations have put Independent Safety Assessment (ISA) certification in place. This produces, in addition to the assessment report, a certificate which indicates that the software has been successfully evaluated. Thus, it is not a product certificate.

11.4.3. *Evolution management*

One of the difficulties with evaluation lies in management of evolutions. Indeed, as is shown by Figure 11.5, a product can be evaluated for various versions, and it is possible for several versions to come and go between two assessments.

Thus, we must be capable of working using the delta method and analyzing the sum of the evolutions. CENELEC 50128:2011 places no constraints at all on the way of working.

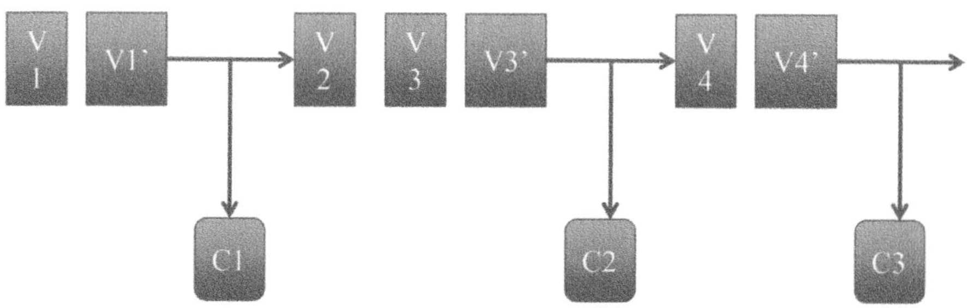

Figure 11.5. *Series of certifications*

11.5. Conclusion

In this chapter, we have presented the issue of assessment and certification. Assessment and certification in the railway domain represent a long and fairly costly process which needs to be prepared as early as possible.

In addition, to conclude, it must be pointed out that successful assessment is linked to skills management, the establishment of a form of organization with appropriate levels of independence, the formalization of the processes and demonstration that these processes have been respected. These processes must cover quality management, development, verification, validation and safety management.

11.6. Appendix: documentation needing to be produced

Document title	Acronym
Assessment	
Assessment plan	AsP
Assessment report	AsR

Conclusion

C.1. Introduction

After a lengthy discussion between the different actors in the railway domain, the 2011 version of CENELEC EN 50128 took account of important issues such as tool qualification and software deployment.

This book presents the different evolutions of the standard, also offering an in-depth explanation of the objectives and impacts.

C.2. CENELEC and IEC

C.2.1. *Compatibility between 50128:2001 and 50128:2011*

Having analyzed the changes that were introduced in the 2011 version of CENELEC 50128, we can conclude that a project evaluated as conforming to the 2011 version also conforms to the 2001 version, but not vice versa.

This point is important because, depending on the country and the different legislation in force, a standard may be cited with or without reference to the version. Unfortunately, in the technical specifications for interoperability (linked to ERTMS), the 2001 version of CENELEC 50128 is cited, while in the SAM regulations in France, CENELEC 50128 is cited without mention of which version (mandating the use of the latest existing version).

C.2.2. *Compatibility between CENELEC 50128 and IEC 62279*

After analyzing the differences between CENELEC 50128:2011 and IEC 62279:2014 [IEC 14], we can conclude that the two are mutually compatible in both directions. IEC 62279:2014 introduces additional points (which are interesting and merit consideration) that do not change the philosophy and/or objectives.

C.3. Changes to come

A draft version of CENELEC 50126 was published in 2012. It includes a complete overhaul of the CENELEC framework (50126, 50129, 50128, 50155) with the aim of establishing a single standard to cover the entirety of the railway system and all parts thereof.

This change is enacted by the establishment of a single standard (CENELEC 50126) to replace CENELEC 50126, 50129 and 50128, maintaining parts of CENELEC 50155, which is devoted to hardware and software.

Part 5 of the new CENELEC 50126 standard will be devoted to software, and should be identical to the 2011 version. In light of the proposals made in the context of IEC 62279, it is very likely that some changes will be made, but these will not have a major impact.

At present, this draft version of the CENELEC 50126 has not been accepted, and a new group has been launched.

One major difficulty with Part 5 of CENELEC 50126 (draft version) will stem from the fact that it applies to any and all types of software, irrespective of the size of the code.

Bibliography

[ABR 96] ABRIAL J.R., *The B-Book*, Cambridge University Press, Cambridge, 1996.

[ADA 01] ADA RESOURCE ASSOCIATION, Operating Procedures for Ada Conformity Assessments, version 3.0, www.ada-auth.org/procs/3.0/ACAP30.pdf, April 2001.

[AFN 07] AFNOR, EN 50155, Railway applications – Electronic equipment used on rolling stock, August 2007

[AFN 90] AFNOR NF F 71-013, Installation fixe et matériel roulant ferroviaires, Informatique, Sûreté de fonctionnement des logiciels – Méthodes appropriées aux analyses de sécurité des logiciels, December 1990.

[AHM 09] AHMED A., LANTERNIER B., "Trois méthodes pour déterminer le niveau de sécurité", *Revue Mesure*, no. 813, pp. 26–27, 2009.

[AIC 93] AMERICAN INSTITUTE OF CHEMICAL ENGINEERS, *Guidelines for Safe Automation of Chemical Processes*, CCPS, New York, 1993.

[ANS 83] ANSI, Norme ANSI/MIL-STD-1815A-1983, Langage de programmation Ada, 1983.

[ARI 92] ARINC, Software Considerations in Airborne Systems and Equipment Certification, DO 178B and by EUROCAE, no. ED12, edition B, 1992.

[BAI 08] BAIER C., KATOEN J.P., *Principles of Model Checking*, The MIT Press, Cambridge, 2008.

[BAR 90] BARANOWSKI F., Définition des objectifs de sécurité dans les transports terrestres, Technical Report 133, INRETS-ESTAS, 1990.

[BAR 03] BARNES J., *High Integrity Software: The SPARK Approach to Safety and Security*, Addison-Wesley, Reading, 2003.

[BAR 14] BARNES J., *Programming in Ada 2012*, Cambridge University Press, 2014.

[BAU 10] BAUFRETON P., BLANQUART J.P., BOULANGER J.L. *et al.*, "Multi-domain comparison of safety standards", *ERTS2*, Toulouse, France, 19-21 May 2010.

[BEH 93] BEHM P., "Application d'une méthode formelle aux logiciels sécuritaires ferroviaires", *Atelier Logiciel Temps Réel, 6es Journées Internationales du Génie Logiciel*, Nantes, France, 1993.

[BEH 96] BEHM P., "Formal development of safety critical software of METEOR", *First B Conference*, Nantes, France, 24-26 November 1996.

[BEN 03] BENVENISTE A., CASPI P., EDWARDS S.A. *et al.*, "The Synchronous Languages 12 Years Later", *Proceedings of the IEEE*, vol. 91, no. 1, January 2003.

[BIE 98] BIED-CHARRETON D., Sécurité intrinsèque et sécurité probabiliste dans les transports terrestres, Technical Report 31, INRETS – ESTAS, November 1998.

[BLA 08] BLAS A., BOULANGER J.L., "Comment améliorer les méthodes d'analyse de risques et l'allocation des THR, SIL et autres objectifs de sécurité", *LambdaMu 16, 16e Congrès de Maîtrise des Risques et de Sûreté de Fonctionnement*, Avignon, France, 6-10 October 2008.

[BOU 99] BOULANGER J.L., DELEBARRE V., NATKIN S., "METEOR: Validation de Spécification par modèle formel", *Revue RTS*, no. 63, pp. 47–62, April-June 1999.

[BOU 00] BOULANGER J.L., GALLARDO M., "Processus de validation basée sur la notion de propriété", *LambdaMu 12*, 28-30 March 2000.

[BOU 03] BOULANGER J.L., "Validation des données liées à la sécurité", *Qualita 2003, Quality and Dependability (RAMS)*, Nancy, France, 19, 20 and 21 March 2003.

[BOU 05] BOULANGER J.L., BERKANI K., "UML et la certification d'application", *ICSSEA 2005*, CNAM, Paris, 1-2 December 2005.

[BOU 06] BOULANGER J.L., Expression et validation des propriétés de sécurité logique et physique pour les systèmes informatiques critiques, Doctoral Thesis, Université de technologie de Compiègne, May 2006.

[BOU 07a] BOULANGER J.L., SCHÖN W., "Assessment of Safety Railway Application", *ESREL*, 2007.

[BOU 07b] BOULANGER J.L., BON P., "BRAIL: d'UML à la méthode B pour modéliser un passage à niveau", *Revue RTS*, no. 95, pp. 147–172, April-June 2007.

[BOU 07c] BOULANGER J.L., "UML et les applications critiques", *Proceedings of Qualita' 07*, Tangier, Morocco, pp. 739–745, 2007.

[BOU 08a] BOULANGER J.L., "RT3-TUCS: How to build a Certifiable and Safety Critical Railway Application", *17th International Conference on Software Engineering and Data Engineering*, SEDE-2008, Los Angeles, pp. 182–187, 30 June-2 July 2008.

[BOU 08b] BOULANGER J.L., GALLARDO M., "Poste de manoeuvre à enclenchement informatique: démonstration de la sécurité", *CIFA, Conférence Internationale Francophone d'Automatique*, Bucharest, Romania, November 2008.

[BOU 09a] BOULANGER J.L., IDANI A., PHILIPPE L., "Linking paradigms in safety critical systems", *Revue ICSA*, September 2009.

[BOU 09b] BOULANGER J.L., "Le domaine ferroviaire, les produits et la certification", *Journée "ligne produit"*, Ecole des Mines de Nantes, 15 October 2009.

[BOU 10a] BOULANGER J.L. (ed.), *Safety of Computer Architectures*, ISTE, London and John Wiley & Sons, New York, 2010.

[BOU 10b] BOULANGER J.L., "Sécurisation des systèmes mécatroniques. Partie 1", *Revue technique de l'ingénieur*, dossier BM 8070, November 2010.

[BOU 11a] BOULANGER J.-L. (ed.), *Static Analysis of Software*, ISTE, London and John Wiley & Sons, New York, 2011.

[BOU 11b] BOULANGER J.L., "Sécurisation des systèmes mécatroniques. Partie 2", *Revue technique de l'ingénieur*, dossier BM 8071, April 2011.

[BOU 11c] BOULANGER J.L. (ed.), *Techniques industrielles de modélisation formelle pour le transport*, Hermès Science-Lavoisier, Paris, 2011.

[BOU 12a] BOULANGER J.-L. (ed.), *Industrial Use of Formal Method: Formal Verification*, ISTE, London and John Wiley & Sons, New York, 2012.

[BOU 12b] BOULANGER J.-L. (ed.), *Formal Methods: Industrial Use from Model to the Code*, ISTE, London and John Wiley & Sons, New York, 2012.

[BOU 13a] BOULANGER J.L. (ed.), *Mise en œuvre de la méthode B*, Hermès Science-Lavoisier, Paris, 2013.

[BOU 13b] BOULANGER J.L., *Safety Management for Software-based Equipment*, ISTE, London and John Wiley, New York, 2013.

[BOU 14a] BOULANGER J.L., BADREAU S., *Ingénierie des exigences – Méthodes et bonnes pratiques pour construire et maintenir un référentiel*, Dunod, Paris, 2014.

[BOU 14b] BOULANGER J.L., ZHAO L., "Sécurité, confidentialité et intégrité dans le domaine ferroviaire", *Lambda Mu 19*, Dijon, France, 21-23 October 2014.

[BOU 14c] BOULANGER JL., *Formal Methods Applied to Industrial Complex Systems: Implementation of the B Method*, ISTE, London and John Wiley, New York, 2014.

[BOW 95] BOWEN J.P., HINCHEY M.G., *Applications of Formal Methods*, Prentice Hall, New York, 1995.

[BRE 00] DE LA BRETESCHE B., *La méthode APTE: Analyse de la valeur, analyse fonctionnelle*, Pétrelle, Paris, 2000.

[CAB 96] MCCABE T.J., WATSON A.H., Structured Testing – A methodology Using the Cyclomatic Complexitry Metric, NIST report, 500-235, 1996.

[CEN 99] CENELEC, NF EN 50126, Railway applications – The specification and demonstration of Reliability, Availability, Maintainability and Safety (RAMS), October 1999.

[CEN 01a] CENELEC, EN 50128, Railway applications – Communications, signaling and processing systems – Software for railway control and protection systems, May 2001.

[CEN 06] CENELEC, EN 50121, Railway applications – Electromagnetic compatibility, 2006.

[CEN 07] CENELEC, Railway Applications – Communication, Signalling and Processing systems – Application Guide for EN 50129 – Part 1: Cross-Acceptance, European norm, May 2007.

[CEN 11] CENELEC, EN 50128, Railway applications – Communications, signaling and processing systems – Software for railway control and protection systems, May 2011.

[CEN 12] CENELEC, 50126, Applications Ferroviaires. Spécification et démonstration de la fiabilité, de la disponibilité, de la maintenabilité et de la sécurité (FMDS), October 2012.

[CHI 94] CHIDAMBER S.R., KEMERER C.F., "A Metric Suite for Object Oriented Design", *IEEE Transactions*, 1994.

[CHO 01] CHOVEAU E., DE CHAZELLES P., "Application de l'ingénierie système à la définition d'une démarche d'ingénierie des exigences pour l'airbus A380", *Génie Logiciel*, no. 59, pp. 13–18, December 2001.

[COU 00] COUSOT P., "Interprétation abstraite", *TSI*, vol. 19, nos. 1–3, www.di.ens.fr/~cousot/COUSOTpapers/TSI00.shtml, 2000.

[COU 77] COUSOT P., COUSOT R., "Abstract interpretation: a unified lattice model for static analysis of programs by construction or approximation of fix points", *Conference Rec. of the 4th Annual ACM SIGPLAN-SIGACT Symp. on Principles of Programming Languages (POPL'77)*, Los Angeles, USA, January 1977, ACM Press, New York, pp. 238–252, 1977.

[DEL 99] DELEBARRE V., GALLARDO M., JUPPEAUX E., NATKIN S., "Validation des constantes de sécurité du pilote automatique de METEOR", *ICSSEA'99*, CNAM, Paris, 1999.

[DIJ 76] DIJKSTRA E.W., *A Discipline of Programming*, Prentice Hall, New York, 1976.

[DO 12] DO 178, Software Considerations in Airborne Systems and Equipment Certification, (Règlementation pour le développement de logiciels dans le secteur aéronautique), version DO-178C, 2012.

[DOR 08] DORMOY F.X., "Scade 6 a model based solution for safety critical software development", *Embedded Real-Time Systems Conference*, Toulouse, France, 2008.

[EIA 98] EIA, EIA-632, Processes for engineering a system, Technical report, April 1998.

[EIA 03] EIA, EIA-632, Processes for engineering a system, Technical report, September 2003.

[GAL 08] GALLARDO M., BOULANGER J.L., "Poste de manœuvre à enclenchement informatique: démonstration de la sécurité", *CIFA, Conférence Internationale Francophone d'Automatique*, Bucharest, Romania, November 2008.

[GAR 94] GARIN H., *AMDEC – AMDE – AEEL – L'essentiel de la méthode*, Editions AFNOR, 1994.

[GEO 90] GEORGES J.P., "Principes et fonctionnement du Système d'Aide à la Conduite, à l'Exploitation et à la Maintenance (SACEM). Application à la ligne A du RER", *Revue générale des chemins de fer*, no. 6, June 1990.

[GUL 04] GULLAND W.G., "Methods of Determining Safety Integrity Level (SIL) Requirements – Pros and Cons", *Proceedings of the Safety-Critical Systems Symposium*, Birmingham, United Kingdom, February 2004.

[HAD 95] HADJ-MABROUCK H., "La maîtrise des risques dans le domaine des automatismes des systèmes de transport guidés : le problème de l'évaluation des analyses préliminaires de risques", *Revue Recherche Transports Sécurité*, no. 49, pp. 101–112, December 1995.

[HAD 98] HADJ-MABROUCK H., STUPARU A., BIED-CHARRETON D., "Exemple de typologie d'accidents dans le domaine des transports guidés", *Revue générale des chemins de fers*, 1998.

[HAD 06] HADDAD S., KORDON F., PETRUCCI L., *Méthodes formelles pour les systèmes répartis et coopératifs*, Hermès Science-Lavoisier, Paris, 2006.

[HAL 91] HALBWACHS N., CASPI P., RAYMOND P., PILAUD D., "The synchronous dataflow programming language Lustre", *Proceedings of the IEEE*, vol. 79, no. 9, pp. 1305–1320, September 1991.

[HAL 05] HALBWACHS N., "A Synchronous Language at Work: the Story of Lustre", *MEMOCODE '05 Proceedings of the 2nd ACM/IEEE International Conference on Formal Methods and Models for Co-Design*, Verona, Italy, 2005.

[HAR 06] HARTWIG K., GRIMM M., MEYER ZU HÖRSTE M., *Tool for the allocation of Safety Integrity Levels*, Level Crossing, Montreal, 2006.

[HAT 94] HATTON L., SAFER C., *Developing Software for High-integrity and Safety-critical Systems*, McGraw-Hill, New York, 1994.

[HOA 69] HOARE C.A.R., "An Axiomatic Basis for Computer Programming", *Communications of the ACM*, vol. 12, no. 10, pp. 576–580, 583, 1969.

[HUL 05] HULL E., JACKSON K., DICK J., *Requirements Engineering*, Springer, Berlin, 2005.

[IDA 06] IDANI A., B/UML: Mise en relation de spécifications B et de descriptions UML pour l'aide ã la validation externe de développements formels en B, Doctoral thesis doctorat, Joseph Fourier University, November 2006.

[IDA 07a] IDANI A., BOULANGER J.L., PHILIPPE L., "A generic process and its tool support towards combining UML and B for safety critical systems", *CAINE 2007*, San Francisco, 7-9 November 2007.

[IDA 07b] IDANI A., OKASA OSSAMI D.D., BOULANGER J.L., "Commandments of UML for safety", *2nd International Conference on Software Engineering Advances IEEE CS*, August 2007.

[IDA 09] IDANI A., BOULANGER J.L., PHILIPPE L., "Linking paradigms in safety critical systems", *Revue ICSA*, September 2009.

[IEC 91] IEC, IEC 1069, Mesure et commande dans les processus industriels – Appréciation des propriétés d'un système en vue de son évolution, Technical report, 1991.

[IEC 02a] IEC 62279, Railway applications – Communications, signaling and processing systems – Software for railway control and protection systems, 2002.

[IEC 02b] IEC 62278, Railway applications – Specification and demonstration of reliability, availability, maintainability and safety (RAMS), 2002.

[IEC 02c] IEC 62280-1, Railway applications – Communication, signaling and processing systems – Part 1: Safety-related communication in closed transmission systems, 2002.

[IEC 02d] IEC 62280-2, Railway applications – Communication, signaling and processing systems – Part 2: Safety-related communication in open transmission systems, 2002.

[IEC 03] IEC, CEI 61131-3:2003, Automates programmables – Partie 3: Langages de programmation, 2003.

[IEC 05] IEC, NF EN 61511, Sécurité fonctionnelle Systèmes instrumentés de sécurité pour le secteur des industries de transformation, European norm, March 2005.

[IEC 06] IEC 60880:2006 Nuclear power plants. Instrumentation and control systems important to safety. Software aspects for computer-based systems performing category A functions, 2006

[IEC 07] IEC 62425, Railway applications – Communication, signaling and processing systems – Safety related electronic systems for signaling, 2007.

[IEC 08] IEC, IEC 61508, Sécurité fonctionnelle des systèmes électriques électroniques programmables relatifs à la sécurité, international norm, 2008.

[IEC 11] IEC, IEC 61513, Centrales nucléaires de puissance – Instrumentation et contrôle-commande importants pour la sûreté – Exigences générales pour les systèmes, August 2011.

[IEC 12] IEC 60571, Railway applications – Electronic equipment used on rolling stock, 2012.

[IEC 14] IEC 62279, Railway applications – Communication, signaling and processing systems – Safety related communication in transmission systems, IEC, 2014

[ISA 05] ISA, Guide d'interprétation et d'application de la norme IEC 61508 et des normes dérivées IEC 61511 (ISA S84.01) et IEC 62061, April 2005.

[ISO 91] ISO 9126:1991, Information technology – Software product evaluation – Quality characteristics and guidelines for their use, 1991.

[ISO 95] ISO/IEC 8652:1995, Information Technology – Programming languages – Ada, 1995.

[ISO 99a] ISO/IEC 18009:1999, Information Technology – Programming Languages – Ada: Conformity Assessment of a Language Processor, 1999.

[ISO 99b] ISO, ISO C standard 1999, Technical report, 1999, www.open-std.org/jtc1/sc22/wg14/www/docs/n1124.pdf.

[ISO 00] ISO/IEC 15942, Technologies de l'information – Langages de programmation – Guide pour l'emploi du langage de programmation Ada dans les systèmes de haute intégrité, 2000.

[ISO 03] ISO/IEC 14882:2003(E), Programming Languages – C++. American National Standards Institute, New York, 2003.

[ISO 04a] ISO, ISO/IEC 90003:2004, Software engineering – Guidelines for the application of ISO 9001:2000 to computer software, 2004.

[ISO 04b] ISO, ISO/IEC 15504-x Information technology – Process assessment, ISO, several parts published between 2004 and 2011.

[ISO 04c] ISO 9126:2004, Information technology – Software product evaluation – Quality characteristics and guidelines for their use, 2004.

[ISO 06] ISO/IEC 14882:2003(E), Programming Languages – C++. American National Standards Institute, New York 10036, 2006.

[ISO 08a] ISO 9000:2008, Quality Management System – Principles and vocabulary, 2008

[ISO 08b] ISO, ISO 9001:2008, Quality Management Systems – Requirements, December 2008.

[ISO 11] ISO, ISO/CD-26262, Road vehicles – Functional safety, 2011.

[ISO 14] ISO 25000, Ingénierie du logiciel – Exigences de qualité du produit logiciel et évaluation (SQuaRE) – Guide de SQuaRE, 2014.

[JON 90] JONES C.B., *Systematic Software Development using VDM*, 2nd edition, Prentice Hall International, Englewood Cliffs, 1990.

[KER 88] KERNIGHAN B.W., RITCHIE D.M., *The C Programming Language*, 2nd edition, Prentice Hall, New York, 1988.

[LAP 92] LAPRIE J.C., AVIZIENIS A., KOPETZ H. (eds), *Dependability: Basic Concepts and Terminology, Dependable Computing and Fault-Tolerant System*, vol. 5, Springer, New York, 1992.

[LER 09] LEROY X., "Formal verification of a Realistic Compiler", *Communication of ACM*, vol. 52, no. 7, pp. 107–115, July 2009.

[LET 00], LE TRUNG B., "Des principes de sécurité GAME, MEM, ALARP", *Recherche Transports Sécurité*, vol. 68, pp. 48–65, July-September 2000.

[LEV 95] LEVESON N.G., *Safeware: System Safety and Computers*, Addison Wesley Publishing Company, Nenlo Park, 1995.

[LIS 90] LISSANDRE M., *Maîtriser SADT*, Armand Colin, Paris, 1990.

[LIS 95] LIS, LABORATOIRE D'INGENIERIE DE LA SURETE DE FONCTIONNEMENT, *Guide de la sûreté de fonctionnement*, Cépaduès, Paris, 1995.

[LOC 05] LOCKHEED M., Joint strike fighter air vehicle C++ coding standards for the system development and demonstration program, Document Number 2RDU00001, Revision C, December 2005.

[MAT 98] MATRA & RATP, "Naissance d'un Métro. Sur la nouvelle ligne 14, les rames METEOR entrent en scène, PARIS découvre son premier métro automatique", *La vie du Rail & des transports*, no. 1076, Hors-série, October 1998.

[MEI 02] MEINADIER J.P., *Le métier d'intégration de systèmes*, Hermès Science-Lavoisier, Paris, 2002.

[MIS 98] MISRA, MISRA-C:1998, Guidelines for the use of the C Language in vehicle based software, April 1998.

[MIS 04] MISRA, MISRA-C:2004, Guidelines for the use of the C language in critical systems, 2004.

[MIS 08] MISRA, MISRA-C++:2008, Guidelines for the use of the C++ language in critical systems, June 2008.

[MIS 12] MISRA, MISRA-C:2012, Guidelines for the use of the C language in critical systems, 2012.

[MON 00] MONIN J.F., *Introduction aux méthodes formelles*, Hermès Science-Lavoisier, Paris, 2000.

[MOR 90] MORGAN C., *Deriving programs from specifications*, Prentice Hall International, Englewood Cliffs, 1990.

[MOT 05] MOTET G., "Vérification de cohérence des modèles UML 2.0", *Première journée thématique Modélisation de Systèmes avec UML, SysML et B-Système*, Association française d'ingénierie système, Toulouse, France, 2005.

[MYE 10] MYERS GLENFORD J., SANDLER C., BADGETT T., *The Art of Software Testing*, Wiley, 3rd edition, December 2011.

[NAU 69] NAUR P., RANDELL B. (eds.), *Software Engineering: A Report on a Conference sponsored by NATO Science Committee*, NATO, 1969.

[OFT 97] OBSERVATOIRE FRANÇAIS DES TECHNIQUES AVANCEES (OFTA), "Applications des Méthodes Formelles au Logiciel", *Arago*, vol. 20, Masson, Paris, June 1997.

[OKA 07] OKALAS OSSAMI D.D., MOTA J.M., THIRY L. *et al.*, "A method to model guidelines for developing railway safety-critical systems with UML", *ICSOFT'07 – International Conference on Software and Data Technologies*, Barcelona, Spain, 2007.

[OMG 07] OMG, Unified Modeling Language (UML), Version 2.1.1, 2007.

[OMG 11] OMG Unified Modeling Language (OMG UML), Infrastructure, OMG, 2011.

[OOT 04a] OOTIA, Handbook for Object-Oriented Technology in Aviation (OOTiA), vol. 1: Handbook Overview, Revision 0, 26 October 2004.

[OOT 04b] OOTIA, Handbook for Object-Oriented Technology in Aviation (OOTiA), vol. 2: Considerations and Issues, Revision 0, 26 October 2004.

[OOT 04c] OOTIA, Handbook for Object-Oriented Technology in Aviation (OOTiA), vol. 3: Best Pratices, Revision 0, 26 October 2004.

[OOT 04d] OOTIA, Handbook for Object-Oriented Technology in Aviation (OOTiA), vol. 4: Certification Practices, Revision 0, 26 October 2004.

[OOT 04d] OOTIA, Handbook for Object-Oriented Technology in Aviation (OOTiA), Certification Practices, Revision 0, vol. 4, 26 October 2004.

[PAP 10] PAPADOPOULOS Y., WALKER M., REISER M.O. *et al.*, "Automatic Allocation of Safety Integrity Levels", *Workshop CARS*, EDCC conference, Valencia, Spain, 2010.

[POH 10] POHL K., *Requirements Engineering: Fundamentals, Principles, and Techniques*, Springer, Berlin, June 2010.

[RAM 09] RAMACHANDRAN M., ATEM DE CARVALHO R., *Handbook of Software Engineering Research and Productivity Technologies: Implications of Globalisation*, Information Science Reference, August 2009.

[RAM 11] RAMACHANDRAN M., *Knowledge Engineering for Software Development LifeCycles: Support Technologies and Applications*, IGI Publishers, April 2011.

[RAS 08] RASSE A., BOULANGER J.L., MARIANO G., THIRY L., PERRONNE J.M., "Approche orientée modèles pour une conception UML certifiée des systèmes logiciels critiques", *CIFA, Conférence Internationale Francophone d'Automatique*, Bucharest, Romania, November 2008.

[RIC 94] RICHARD-FOY M., LEGOFF G., "On-board with safety critical software: Implementing safety critical software for high-speed railway transportation", *Alsys World Dialogue*, vol. 18, no. 2, 1994.

[ROQ 06] ROQUES P., *UML 2 par la pratique – Etudes de cas et exercices corrigés*, Eyrolles, Paris, 2006.

[ROQ 07] ROQUES P., *UML 2 – Modéliser une application Web*, Eyrolles, Paris, 2007.

[RSS 07] RAIL SAFETY AND STANDARDS BOARD, *Engineering Safety Management (The Yellow Book) – Fundamentals and Guidance*, vol. 1–2, no. 4, 2007.

[RTA 11a] RTA DO-178:C, Software consideration in airborne systems and equipement certification, version C, December 2011.

[RTA 11b] RTA DO 330, Software Tools Qualification Consideration, version C, December 2011.

[SCO 98] MEYERS S., *Effective C++: 50 Specific Ways to Improve Your Programs and Design*, 2nd edition, Addison-Wesley, Reading, 1998.

[SIM 06] SIMON C., SALLAK M., AUBRY J.F., "Allocation de SIL par agrégation d'avis d'experts", *LambdaMu 15*, 2006.

[SMI 07] SMITH D.J., SIMPSON K.G.L., *Functional Safety, A Straightforward Guide to Applying IEC 61508 and Related Standards*, 2nd edition, Elsevier, Engelska, 2007.

[SOM 07] SOMMERVILLE I., Software Engineering, version 8, 2007.

[SPI 89] SPIVEY J.M., *The Z Notation – A Reference Manual*, Prentice Hall International, London, 1989.

[STA 94] THE STANDISH GROUP, The chaos report, Technical report, 1994.

[STA 01] THE STANDISH GROUP, Extreme chaos, Technical report, 2001.

[SUT 04] SUTTER H., ALEXANDRESCU A., *C++ Coding Standards, 101 Rules, Guideline and Best Practices*, Addison-Wesley, Reading, 2005.

[UNI 09] UNIFE, International Railway Industry Standard IRIS, revision 02, 2009.

[VIL 88] VILLEMEUR A., *Sûreté de fonctionnement des systèmes industriels*, Eyrolles, Paris, 1988.

[XAN 00] XANTHAKIS S., REGNIER P., KARAPOULIOS C., *Le test des logiciels*, Hermès Science, Paris, 2000.

Glossary

AAD	Application Architecture and Design
ACATS	Ada Conformity Assessment Test Suite
AFIS	*Association Française d'Ingénierie Système* (French Systems Engineering Association)
AITR	Application Integration Test Report
AITS	Application Integration Test Specification
ANSI	American National Standards Institute
AP	Auto-Pilot
APP	Application Preparation Plan
APVR	Application Preparation Verification Report
ARS	Application Requirements Specification
ASIL	Automotive Safety Integrity Level
ASN	*Autorité de Sûreté Nucléaire* (French Nuclear Safety Authority)
ASR	Assessor
ATR	Application Test Report
ATS	Application Test Specification
AVR	Application data/algorithm Verification Report
CARE	Common Airbus Requirements Engineering
CAS	Computer-Aided Specification

CATR	Complete Application Test Report
CATS	Complete Application Test Specification
CBTC	Communication-Based Train Control
CCR	Critical Code Reading
CDG	Charles de Gaulle
CENELEC[1]	*Comité Européen de Normalisation ELECtrotechnique* (European Committee for Electrotechnical Standardization)
CMMi	Capability Maturity Model for integration
CMP	Configuration Management Plan
COFRAC	*COmité FRançais d'Accréditation* (French Accreditation Committee)
COTS	Commercial off-the-shelf
CTR	Component Tests Report
CTs	Component Tests
CTS	Component Tests Specification
CV	Curriculum Vitae
DAL	Design Assurance Level
DES	Designer
DoD	Department of Defense
DRBCS	Dynamic Radio-Based Control System
DVR	Data Verification Report
E/E/PE	Electrical/Electronic/Programmable Electronics
EMC	Electromagnetic Compatibility
ESP	Electronic Stability Program
FMEA	Failure Mode and Effect Analysis
FR	Failure Rate
FTA	Fault Tree Analysis
FTs	Functional Tests

1 For further information, see: www.cenelec.eu.

GAME	*Globalement Au Moins Equivalent* (Globally At Least Equivalent)
HAZOP	HAZard and OPerability study
HL	Hazard Log
HLR	High Level Requirement
HR	Highly Recommended
H/S	Hardware/Software
HSITR	Hardware/Software Integration Test Report
HSITS	Hardware/Software Integration Test Specification
IdM	*Ingénierie des Modèles* (Model-based Engineering)
IEC[2]	International Electrotechnical Commission
IMP	IMPlementer
INCOSE[3]	International Council on Systems Engineering
INT	INTegrator
IRIS[4]	International Railway Industry Standard
ISA	Independent Safety Assessor
ISO[5]	International Standardization Organization
ITR	Installation Test Report
ITS	Installation Test Specification
ITs	Integration Tests (Chapter 5)
KBL	Known Bug List
LLR	Low Level Requirement
LOPA	Layer of Protection Analysis
M	Mandatory
MBD	Model-Based Development

2 See: www.iec.ch/.

3 See: www.incose.com

4 For further information, see: www.iris-rail.org.

5 See: www.iso.org/iso/home.htm.

METEOR	*METro Est-Ouest Rapide* (rapid transit East/West Line of Paris Metro)
MISRA[6]	Motor Industry Software Reliability Association
nOOm	*n* Out Of *m* architecture
NR	Not Recommended
OATR	Overall Application Test Report
OCC	Operation Control Center
OMG[7]	Object Management Group
OO	Object Oriented
OOTiA[8]	Object Oriented Technology in Aviation
OSTR	Overall Software Tests Report
OSTS	Overall Software Tests Specification
OTs	Overall Tests
PAING	*Poste Aiguillage Informatisé – Nouvelle Génération* (Computerized Points System – New Generation)
PDS	Preliminary Design Specification
PHA	Preliminary Hazard Analysis
PIPC	*Poste Informatique Petite Capacité* (Small Capacity Computer Station)
PK	*Point Kilométrique* (Kilometric Point)
PLCs	Programmable logic controllers
PM	Project Manager
PMI	*Poste de Manœuvre Informatisé* (Computerized Operation Station)
PQP	Project Quality Plan
PSC	*Processeur Sécuritaire Codé* (Coded Safety Processor)
QAM	Quality Assurance Manual
QMS	Quality Management System

6 See: www.misra.org.uk/.

7 See: www.omg.org/.

8 See: www.faa.gov/aircraft/air_cert/design_approvals/air_ software/oot/.

QUA	QUAlity team
QUAM	Quality Assurance Manager
QVR	Quality Verification Report
R	Recommended
RAM	Reliability, Availability, Maintainability
RAMS	Reliability, Availability, Maintainability and Safety
RATP[9]	*Régie Autonome des Transports Parisiens* (Parisian Transport Operator)
REQ	REQuirement
RER	*Réseau Express Régional* (Parisian Suburban Rail Network)
RQM	ReQuirements Manager
S/S	Software/Software
SACEM	*Système d'Aide à la Conduite, à l'Exploitation et à la Maintenance* (Driving, Operation and Maintenance Support System)
SADT	Structured Analysis and Design Technique
SADVR	Software Architecture and Design Verification Report
SAET	*Système d'Automatisation de l'Exploitation des Trains* (Automatican System for Trains)
SAP	Safety Assurance Plan
SAS	Software Architecture Specification
SC	Safety Case
SCADE	Safety Critical Application Development Environment
SCDVR	Software Component Design Verification Report
SCMP	Software Configuration Management Plan
SCS	Software Component Specification
SCTS	Software Component Test Specification
SDS	Software Design Specification

9 For further information, see: www.ratp.fr/.

SEEA	Software Error Effects Analysis
SIL	Safety Integrity Level
SIS	Software Interface Specifications
SITR	Software Integration Test Report
SITS	Software Integration Test Specification
SIVR	Software Integration Verification Report
SPICE	Software Process Improvement and Capability dEtermination
SPIN	SeParations and INcineration
SQAP	Software Quality Assurance Plan
SQuaRE	Software Quality Requirements and Evaluation
SSAP	Software Safety Assurance Plan
SSIL	Software Safety Integrity Level
SVaP	Software Validation Plan
SVeP	Software Verification Plan
SVR	Software Validation Report
SwRS	Software Requirement Specification
SwRSVR	Software Requirements Verification Report
SwVS	Software Version Sheet
SyRS	System Requirement Specification
SysML	System Modeling Language
SyVS	System Version Sheet
TC	Track Circuit
TCL	Tool Confidence Level
TCMS	Train Control/Management System
TCO	*Tableau de Contrôle Optique* (Optical Control View)
TCs	Tests Cases
THR	Tolerable Hazard Rate
TQR	Tools Qualification Report

TSN	Technical System Needs
TVM	*Transmission Voie Machine* (Track-to-Machine Transmission)
TVR	Tool Validation Report
Tx	Objective of qualification of a tool: T1, T2 or T3
UML	Unified Modeling Language
US	*Unité de Surveillance* (Monitor Unit)
UTs	Unit Tests
V&V	Verification and Validation
VAL	VALidator
VER	VERifier
VVP	Verification and Validation Plan

Index

O, P, Q

R, S

T, U

V, W

Other titles from

iSTE

in

Control, Systems and Industrial Engineering

2014

DAVIM J. Paulo
Machinability of Advanced Materials

ESTAMPE Dominique
Supply Chain Performance and Evaluation Models

FAVRE Bernard
Introduction to Sustainable Transports

MICOUIN Patrice
Model Based Systems Engineering: Fundamentals and Methods

MILLOT Patrick
Designing Human–Machine Cooperation Systems

MILLOT Patrick
Risk Management in Life-Critical Systems

NI Zhenjiang, PACORET Céline, BENOSMAN Ryad, REGNIER Stéphane
Haptic Feedback Teleoperation of Optical Tweezers

OUSTALOUP Alain
Diversity and Non-integer Differentiation for System Dynamics

REZG Nidhal, DELLAGI Sofien, KHATAD Abdelhakim
Joint Optimization of Maintenance and Production Policies

STEFANOIU Dan, BORNE Pierre, POPESCU Dumitru, FILIP Florin Gh., EL KAMEL Abdelkader
Optimization in Engineering Sciences: Metaheuristics, Stochastic Methods and Decision Support

2013

ALAZARD Daniel
Reverse Engineering in Control Design

ARIOUI Hichem, NEHAOUA Lamri
Driving Simulation

CHADLI Mohammed, COPPIER Hervé
Command-control for Real-time Systems

DAAFOUZ Jamal, TARBOURIECH Sophie, SIGALOTTI Mario
Hybrid Systems with Constraints

FEYEL Philippe
Loop-shaping Robust Control

FLAUS Jean-Marie
Risk Analysis: Socio-technical and Industrial Systems

FRIBOURG Laurent, SOULAT Romain
Control of Switching Systems by Invariance Analysis: Application to Power Electronics

GRUNN Emmanuel, PHAM Anh Tuan
Modeling of Complex Systems: Application to Aeronautical Dynamics

HABIB Maki K., DAVIM J. Paulo
Interdisciplinary Mechatronics: Engineering Science and Research Development

HAMMADI Slim, KSOURI Mekki
Multimodal Transport Systems

JARBOUI Bassem, SIARRY Patrick, TEGHEM Jacques
Metaheuristics for Production Scheduling

KIRILLOV Oleg N., PELINOVSKY Dmitry E.
Nonlinear Physical Systems

LE Vu Tuan Hieu, STOICA Cristina, ALAMO Teodoro,
CAMACHO Eduardo F., DUMUR Didier
Zonotopes: From Guaranteed State-estimation to Control

MACHADO Carolina, DAVIM J. Paulo
Management and Engineering Innovation

MORANA Joëlle
Sustainable Supply Chain Management

SANDOU Guillaume
Metaheuristic Optimization for the Design of Automatic Control Laws

STOICAN Florin, OLARU Sorin
Set-theoretic Fault Detection in Multisensor Systems

2012

AÏT-KADI Daoud, CHOUINARD Marc, MARCOTTE Suzanne, RIOPEL Diane
Sustainable Reverse Logistics Network: Engineering and Management

BORNE Pierre, POPESCU Dumitru, FILIP Florin G., STEFANOIU Dan
Optimization in Engineering Sciences: Exact Methods

CHADLI Mohammed, BORNE Pierre
Multiple Models Approach in Automation: Takagi-Sugeno Fuzzy Systems

DAVIM J. Paulo
Lasers in Manufacturing

DECLERCK Philippe
Discrete Event Systems in Dioid Algebra and Conventional Algebra

DOUMIATI Moustapha, CHARARA Ali, VICTORINO Alessandro,
LECHNER Daniel
Vehicle Dynamics Estimation using Kalman Filtering: Experimental Validation

HAMMADI Slim, KSOURI Mekki
Advanced Mobility and Transport Engineering

MAILLARD Pierre
Competitive Quality Strategies

MATTA Nada, VANDENBOOMGAERDE Yves, ARLAT Jean
Supervision and Safety of Complex Systems

POLER Raul *et al.*
Intelligent Non-hierarchical Manufacturing Networks

YALAOUI Alice, CHEHADE Hicham, YALAOUI Farouk, AMODEO Lionel
Optimization of Logistics

ZELM Martin *et al.*
I-EASA12

2011

CANTOT Pascal, LUZEAUX Dominique
Simulation and Modeling of Systems of Systems

DAVIM J. Paulo
Mechatronics

DAVIM J. Paulo
Wood Machining

KOLSKI Christophe
Human-computer Interactions in Transport

LUZEAUX Dominique, RUAULT Jean-René, WIPPLER Jean-Luc
Complex Systems and Systems of Systems Engineering

ZELM Martin, *et al.*
Enterprise Interoperability: IWEI2011 Proceedings

2010

BOTTA-GENOULAZ Valérie, CAMPAGNE Jean-Pierre, LLERENA Daniel,
PELLEGRIN Claude
Supply Chain Performance / Collaboration, Alignement and Coordination

BOURLÈS Henri, GODFREY K.C. Kwan
Linear Systems

BOURRIÈRES Jean-Paul
Proceedings of CEISIE'09

DAVIM J. Paulo
Sustainable Manufacturing

GIORDANO Max, MATHIEU Luc, VILLENEUVE François
Product Life-Cycle Management / Geometric Variations

LUZEAUX Dominique, RUAULT Jean-René
Systems of Systems

VILLENEUVE François, MATHIEU Luc
Geometric Tolerancing of Products

2009

DIAZ Michel
Petri Nets / Fundamental Models, Verification and Applications

OZEL Tugrul, DAVIM J. Paulo
Intelligent Machining

2008

ARTIGUES Christian, DEMASSEY Sophie, NÉRON Emmanuel
Resources–Constrained Project Scheduling

BILLAUT Jean-Charles, MOUKRIM Aziz, SANLAVILLE Eric
Flexibility and Robustness in Scheduling

DOCHAIN Denis
Bioprocess Control

LOPEZ Pierre, ROUBELLAT François
Production Scheduling

THIERRY Caroline, THOMAS André, BEL Gérard
Supply Chain Simulation and Management

2007

DE LARMINAT Philippe
Analysis and Control of Linear Systems

LAMNABHI Françoise *et al.*
Taming Heterogeneity and Complexity of Embedded Control

LIMNIOS Nikolaos
Fault Trees

2006

NAJIM Kaddour
Control of Continuous Linear Systems

Printed and bound by CPI Group (UK) Ltd, Croydon, CR0 4YY
13/09/2026
14961078-0005